SYMPATHETIC VIBRATIONS

SYMPATHETIC VIBRATIONS

REFLECTIONS ON PHYSICS AS A WAY OF LIFE

K. C. COLE

foreword by
Frank Oppenheimer

William Morrow and Company, Inc.
New York

Some of the material in these chapters has been adapted from essays published by the author in *The New York Times*, *The Washington Post*, and *Discover*.

Grateful acknowledgment is made for permission to reprint the following:

Photographs on pages 26, 31, 41, 48, 56, 59, 60, 61, 64, 70, 72, 84, 98, 121, 125, 127, 153, 221, 233, 235, 236, 237, 238, 239, 240, 241, 244, 246, 247, 253, 255, 257, 258, 281, 284, 288, 292, 293, 317, and 321 courtesy of Nancy Rodger.

Illustrations on pages 45, 49, 65, 155, 166, 167, 209, 234, 251, 263, 269, 270, 294 courtesy of The Exploratorium.

Illustrations on pages 28, 83, 86, 87, 89, 92, 101, 118, 141, 147, 148, 151, 212, and 286 (bottom) reprinted from *Conceptual Physics* (Fourth Edition) by Paul G. Hewitt, 1981, courtesy of Little, Brown and Company.

Illustrations on pages 94, 105, 111, 133, 158, 219, 227, 315, 327 reprinted from *Knowledge and Wonder: The Natural World as Man Knows It* (Second Edition) by Victor F. Weisskopf. Copyright © 1962, 1966 by Doubleday & Company, Inc. Reproduced by permission of the publisher.

Photographs on pages 84, 242, 243, 245, 252, 277, 278, 329, and 337 by Susan Schwartzenberg.

Photographs on pages 42, 44, 53, 142, and 174 courtesy of the Stanford Linear Accelerator Center.

Illustrations on pages 43, 100, 113, 144, and 172 from *Biography of Physics*, copyright © 1961 by George Gamow. Reprinted by permission of Harper & Row, Publishers, Inc.

Illustrations on pages 103, 107, 110, 114, 152 and 286 (top) from *Music of the Spheres: The Material Universe—From Atom to Quasar, Simply Explained* by Guy Murchie. Copyright © 1961 by Guy Murchie. Reprinted by permission of Houghton Mifflin Company.

For my daughter,
Elizabeth

Foreword

I first met K. C. Cole when, as associate editor for the *Saturday Review*, she wrote and published an article about The Exploratorium. That was November 1972.

Since then she has been learning and writing about perception and physics. She has played with, talked about, and written about the exhibits at The Exploratorium. She has watched the world in her kitchen, in airplanes, in automobiles, on sailboats, and in swimming. Everywhere she finds something new, something new that turns out to be connected with everything else that she knew.

But K.C. has also plunged deep into the world of modern physics. She has talked with physicists in Pasadena, Palo Alto, Berkeley, New York City, Cambridge, Ithaca, Brookhaven, Murray Hill, Princeton. And there are a lot of good physicists, clear explainers, and thoughtful idealists in that collection of towns.

In learning physics in this way, she first began to notice more of what is going on around her. But to help in the noticing and to help tie all the different observed effects together, she also discovered that physicists (and of course all scientists) have learned that there are special ways of thinking about nature that turn out to be crucially helpful, not only in tying together what is already known but also in leading one on to previously undiscovered marvels. These ways of thinking about nature and natural phenomena can be as simple as precise definitions of a word. I remember realizing in my high school physics course how much further my understanding had advanced merely by defining *pressure* as a *force per unit* area. Suddenly it became clear to me how hydraulic jacks and elevators could lift large masses with the application of only small forces. What a magic word that "per" turned out to be. Miles per hour, electric charge per volt, electric energy per second . . . and on and on. Recently I have had the chance to observe the effect that the word "no" can have on a broad range of situations that confront a ten-month-old infant. Throughout my own life there have been innumerable times when the careful definition of words has helped me to solve moral or emotional problems.

In other instances the "way of thinking" is more nearly a visualization. Such is the case with the interference of waves. How otherwise can one understand how two energetic light waves falling on the same place (in one's retina or on a photograph) can produce no effect, how two lights can make darkness. One has to think about waves in a new way in order to do so, but once this way of thinking has been found, it can then be applied to sound and water and light, as well as to electrons and probability waves. But does it also apply to fashion waves or economic waves? Why not? If one has a new way of thinking, why not apply it wherever one's thoughts lead to? It is certainly entertaining to let oneself do so, but it is also often very illuminating and capable of leading to new and deep insights.

Physical science used to be called Natural Philosophy, but unfortunately physics is no longer taught as a course in philosophy. This situation is surprising because most of the physicists that I know talk with me and with each other with full awareness that the way in which we think of the physical world profoundly shapes the way we think of the human and ethical worlds. For them physics is a part of culture and of philosophy. In the chapters of this book, K. C. Cole has been able to talk to non-physicists in the thoughtful way that many physicists talk to each other.

In addition to talking to physicists, K. C. Cole has read extensively what Bohr and Einstein and Weisskopf and my brother and many others have written about the philosophic implications of modern physics. She has therefore been stimulated to let her own thoughts wander and to productively carry the imagery of physics into human affairs and feelings. Some scientists frown on this practice of carrying the sharply defined ways of thinking about physics into the more loosely defined domain of human feelings and actions. Sometimes this disapproval is mere cowardice; they fear getting into the quicksand of human relationships. Sometimes their resistance is due to a kind of clubbiness, a hiding behind the protection of their professional jargon. Sometimes it is because,

although they know better than to do so, they think others will not
be as wise and will take the analogies too seriously. They are quite
willing to talk about how the exchange of virtual photons between
two electrons can account for the attraction or repulsion of these
charges, but they are much too timid and unsure of themselves to
point out that the exchange of unspoken words between two people can account for their mutual attractions and repulsions.

Fortunately K. C. Cole has no such fears or timidities. As she
learns new ways of thinking about matter and energy and light and
sound and motion, fundamental particles and symmetries, she is
quite wonderfully willing to let us participate in her delightful ap-
plications of these imageries to the everyday world of the people
and things around her . . . and us.

> —FRANK OPPENHEIMER
> Director of The Exploratorium,
> a museum of science and
> perception in San Francisco

Acknowledgments

My special thanks to Phyllis Theroux, for first suggesting that I try writing about the unlikely subject of physics as a way of life in four essays for the *Washington Post*; to Nancy Newhouse of *The New York Times*, for encouraging me to pursue the same subject in a series of seven "Hers" columns; to Maria Guarnaschelli, my editor at William Morrow, for her enthusiasm and good suggestions; to Leon Jaroff, editor of *Discover* magazine, for faith, flexibility, and a column in his magazine in which to reflect on the many facets of "natural philosophy"; to Frank Oppenheimer and Viki Weisskopf, for inspiration and above all for the time and care that went into correcting and clarifying these ideas; to Nancy Rodger for her illuminating photographs, and to all The Exploratorium staff; and finally to Delphi Callejas, who looked after my family while I looked after my book.

CONTENTS

Contents

Forces and Influences

The ideas in this book were molded largely from the works and thoughts of others—from the ideas and opinions of Albert Einstein to the writings of people like Guy Murchie and Lincoln Barnett and the always original insights of "my friend the physicist." Some of these people appear so often on the following pages that it seems only fitting to properly introduce them here:

Frank Oppenheimer, director of The Exploratorium in San Francisco, is my friend the physicist, and has been since I first stumbled upon his amazing museum almost twelve years ago. I often think that The Exploratorium resembles nothing so much as the inside of Frank's brain: a rather chaotic but deeply connected combination of art, science, philosophy, education, politics, and pure play. The Exploratorium is widely known as the world's best science museum, but in fact it is a museum of human perception. Since the beginning, a battered sign on the shop has announced: HERE IS BEING CREATED AN EXPLORATORIUM—A COMMUNITY MUSEUM DEDICATED TO HUMAN AWARENESS. Human awareness has always been Frank's primary concern, since his days wandering New York City's water towers late at night and writing essays on the view from the top, to his student experiences at Johns Hopkins and Caltech practicing physics and the flute, through the years he worked on the atomic bomb project at Los Alamos to his adventures as cosmic ray physicist, to ten years of virtual political exile (the price he paid for his pacifist views) and cattle ranching in the Colorado Rockies, to his return to teaching, and finally the culmination of all his experiences in The Exploratorium. Frank has received a great number of honors for his work, including two Guggenheim Fellowships and distinguished service awards from Caltech, the American Association of Physics Teachers, and the American Association of Museums.

Victor Weisskopf should really be called "my other friend the physicist." I first met him through his excellent book *Knowledge and Wonder*, which is the only book other than *The Exorcist* I have stayed up all night to read. (I later discovered his *Physics in the 20th Century*, which is equally rich.) Weisskopf is Institute Professor emeritus at M.I.T., National Medal of Science winner, director of CERN (the European Organization for Nuclear Research) during its most important formative years, former president of the American Academy of Arts and Sciences, and current member of the Pontifical Academy of Sciences, where he is deeply involved in issues of nuclear disarmament. Despite all this, he is dedicated to the importance of what some scientists still scorn as "popularization" and always tries to make time to work with science writers and explain the central ideas of physics to the general public. Weisskopf, like Oppenheimer, is remarkably and refreshingly relaxed about other people using his ideas. He often says: "The only sin is if you hear a good idea and you *don't* use it."

Philip Morrison, Institute Professor at M.I.T., astrophysicist, author of many books (most recently *Powers of Ten* with his wife, Phylis), and book reviewer for *Scientific American*, is probably best known as a spellbinding lecturer. His approach is always original, and he is a master at turning the obvious inside out and forcing you to reevaluate your assumptions. On the several occasions we have talked at M.I.T. and at the Oppenheimers' house in San Francisco, he has left me both marveling at my own silliness and filled to the brim with new ideas.

Richard Feynman was once described as "the smartest man in the world." True or not, he is certainly one of the most colorful characters in the physics community. A Nobel Prize winner for his work in quantum electrodynamics, accomplished player of bongo drums, and creator of the famous Feynman diagrams for visually representing subatomic events, he is currently a theoretical physicist at Caltech—where his Brooklyn accent is amusingly out of place. I once lucked into an hour-long talk with Feynman (he is notoriously press shy), but much of his material presented here is from the indispensable *Feynman Lectures in Physics* and also his book *The Character of Physical Law.*

Guy Murchie is a well-known science writer and author of *Song of the Sky, Music of the Spheres,* and *The Seven Mysteries of Life.* His books overflow with facts, quotes, anecdotes, and most of all enthusiasm for all things natural. I take great pride in having introduced Guy Murchie to The Exploratorium—and vice versa.

Albert Einstein, of course, was the universally acclaimed genius who crystallized both the Theory of Special Relativity ($E = mc^2$, time dilation, and all that) and the Theory of General Relativity (curved space, black holes, and all that). He revolutionized the way people think about time, space, matter, energy, motion, and other fundamental phenomena. But far more than a brilliant scientist, Einstein was a great humanist who talked, wrote, and worried about war, the human condition, tyranny, and most of all about the proliferation of nuclear bombs—something, he once said, that has changed everything but our way of thinking.

J. Robert Oppenheimer, Frank's older brother, is often credited with developing the first school of theoretical physics in the United States after his return from Europe in the 1920s. As the scientist in charge of Los Alamos, he is often known by the unfortunate title father of the atomic bomb. Still, he was primarily known as a great teacher, deep thinker, and, in the end, something of a martyr for his political views: he was stripped of his security clearance during the McCarthy purges of the 1950s, partly because of his opposition to Edward Teller's determination to produce the vastly more powerful hydrogen bomb.

George Gamow was an eccentric and important physicist who was one of the first major contributors to the now almost universally accepted Big Bang Theory of the origins of the universe.* His enchanting stories of how a poor bank clerk came to learn about relativity and quantum physics (*Mr. Tompkins in Wonderland*

---

*Actually, what's universally accepted is rapid expansion of the universe in all directions—something which implies that at one time it was all compressed together at one point.

and *Mr. Tompkins Explores the Atom*) are a delightful introduction for anyone interested in modern science. I also highly recommend his *One, Two, Three . . . Infinity, Biography of Physics*, and *Biography of the Earth*.

Stephen Jay Gould is the outspoken and original Harvard biologist and geologist who writes those wonderful essays for *Natural History* magazine, collected in *Ever Since Darwin, The Panda's Thumb*, and *Hen's Teeth and Horse's Toes*. Gould rarely makes a point about evolutionary biology without also drawing a parallel to broader aspects of human affairs.

Vera Kistiakowsky is an experimental physicist and professor at M.I.T. who has taken the time on several occasions to review my various writings and to talk with me about physics.

Sir James Jeans was a British astronomer and physicist whose research covered a wide range from molecular physics to quantum theory and cosmology. He ceased research in 1928 (after he was knighted) to popularize science. His lectures and radio speeches were published in *The Universe Around Us* and *The Mysterious Universe*.

Sir Arthur Eddington was a contemporary of Jeans's (the two disagreed on aspects of both astronomy and philosophy) who specialized in relativity theory. In fact, Eddington was the first to interpret Einstein's theory of relativity in English. Einstein considered Eddington's presentation the finest in any language.

Niels Bohr was a Danish physicist who is widely known as the father of quantum mechanics: he inspired a whole generation of physicists with his ideas about science and its implications in human thought. Bohr was the first to attribute the specific properties of atoms to the fact that events within an atom (like the emission of light) happen only as a whole—a quantum leap, so to speak. He also developed the idea of complementary descriptions to reconcile this strangely particlelike aspect of radiation with its wavelike character.

Isaac Newton was the seventeenth-century scientist who is pictured by most people as sitting under a tree waiting for an apple to fall on his head. True or apocryphal, it was Newton who first saw that the fall of the apple and the orbit (or "fall") of the moon were propelled by the same force—gravity. His three famous laws of motion (every action has an equal and opposite reaction, and so on) are familiar to every schoolchild. He created "the calculus," first realized that white light was actually a mixture of the full spectrum of colors, and in the end was obsessed with alchemy and mysticism.

Aristotle was an ancient Greek philosopher whose image of the universe as static, finite, and earth-centered dominated scientific thinking on and off for two thousand years. Aristotle's primary contributions were not in physics, but he is often credited (or rather blamed) for getting future scientists off on the wrong tack—especially with his laws of motion which erroneously assumed that the "natural" state of a body was at rest and that things would naturally come to a stop if they were no longer being pushed by a force.

Nicholas Copernicus was the fifteenth-century astronomer who is credited with discovering the fact that the earth moves around the sun and not vice versa. Many historical accounts say that the new Copernican system greatly simplified the old system developed by the Greek philosopher Ptolemy, which saw the motions of the planets, sun, and moon as a complicated collection of circles within circles—or epicycles. But Copernicus's system also required many complicated motions—primarily because he still saw the motions of heavenly bodies as perfect circles, even though the true orbits of the planets are ellipses.

Galileo Galilei was the sixteenth- and seventeenth-century scientist who is credited with bringing back the importance of experimental evidence into the study of physical phenomena. Galileo is often pictured standing at the top of the leaning tower of Pisa—dropping a feather and a rock to prove that they fall at the same rate (an impossibility, unless Pisa existed in a vacuum). He is also credited with noticing that the period of a pendulum depends only on the length of the pendulum and not on the size of the swing; for inventing telescopes; for discovering mountains on the moon, numerous new stars, and the moons of Jupiter—the first evidence that planets other than our own could have "satellites."

Johannes Kepler was a contemporary of Galileo's who spent his life searching for cosmic harmonies in the "music of the spheres." Probably his most important contribution was his recognition that the planets were held in orbit around the sun by a *force;* he also developed the three famous laws of planetary motion, which finally proved that the orbits of the planets were ellipses and not circles.

Of course, this is only a partial list, organized more or less by frequency of appearance. It would be all but impossible to list all of the forces and influences that shaped this book. However, I do want to add the following authors as comprising what I would call an essential bibliography:

Isaac Asimov, *On Physics, On Chemistry;* Adolph Baker, *Modern Physics and Antiphysics;* Lincoln Barnett, *The Universe and Dr. Einstein;* Max Born, *The Natural Philosophy of Cause and Effect;* J. Bronowski, *The Ascent of Man, Science and Human Values;* Annie Dillard, *Pilgrim at Tinker Creek;* Loren Eiseley, *The Unexpected Universe;* A. P. French, editor, *Einstein: A Centenary Volume;* R. L. Gregory, *The Intelligent Eye;* Paul G. Hewitt, *Conceptual Physics;* Sir James Jeans, *Physics and Philosophy;* Daniel J. Kevles, *The Physicists: The History of a Scientific Community in Modern America;* Arthur Koestler, *The Sleepwalkers;* P. B. Medawar, *Advice to a Young Scientist;* B. K. Ridley, *Time, Space and Things;* Carl Sagan, *Cosmos;* Peter S. Stevens, *Patterns in Nature;* James Trefil, *The Unexpected Vista;* Judith Wechsler, editor, *On Aesthetics in Science.*

PHYSICS AS A WAY OF LIFE

Does everything around you seem to be falling apart? (In physics, the term used to describe the increasing disorder in the universe is entropy.) Does it seem to take a monumental effort to get going on a new task, or break the bonds of old habits? (Your problem might have something to do with inertia.) Does time seem sometimes to fly? At other times to stand still? (Time is actually a lot more flexible than most of us think—both psychologically *and* physically.) Do you feel at times totally in tune with your job, your friends, your generation? At other times irritatingly discordant? (One way to look at it is the physical principle of resonance.) Does it frustrate you when the same event or conversation is interpreted in different ways by different people? (Perception is in the eye of the beholder.) Do you find yourself agreeing with both sides of irreconcilable arguments? (Complementarity also explains how light can be both waves and particles.)

It seems absurd, of course, to apply abstract scientific principles to the personal concerns of everyday life. The inertia of a planet or a proton is not at all the same inertia that relegates people to psychological ruts. Yet the analogies between physics and human affairs are already part and parcel of our everyday thoughts and language. We speak of people who attract and repel each other (like magnets), of the force of habit, of cause and effect, of disorder and quantum leaps, most of all of space and time. Our language is liberally sprinkled with the metaphors of science—and the language of science is inescapably infused with images from everyday life. If the behavior of atoms and people seems sometimes to share the same forms, it is not so surprising. It seems strange only until you stop to think about it: after all, we are all held together (and apart) by the same forces; we are subject to the same natural laws; we are fashioned from the same basic stuff.

Not all scientists believe that analogies between their sharply defined discipline and the far more murky area of human affairs are possible; everyone insists that they be approached with the utmost care. If this book is heavy with quotes, it is partly an admission that I know I am treading on very delicate ground, and that it needs to be shored up with the strongest possible arguments every step of the way.

Yet just as many scientists think that science and the humanities have suffered a painful and unnatural separation. Far more than a collection of facts, science is a body of ideas that forms the cultural context through which we view the world. It influences, and is influenced by, almost everything else around us—from the role of religion to the status of slaves. Science started out as "natural philosophy." When the so-called scientific revolution of the seventeenth century produced such signal works as Kepler's *Harmony of the World* and Galileo's *Message from the Stars*, the new discoveries were considered part of what was then called the New Philosophy. This, too, should not surprise us. Both philosopher and physicist are concerned with the causes of things, with the questions Why are things the way they are? and Why do they behave the way they do?

Lately, scientists have largely lost their status as philosophers (although they may be fast regaining it through the efforts of such philosopher/scientists as Lewis Thomas and Stephen Jay Gould). Many people think of science primarily in terms of technology or as the stuff of esoteric equations. Yet history shows that science has long been a shaper of human thought. As Nobel Prize winner Max Born put it: "It has been said that the metaphysics of any period is the offspring of the physics of the preceding period. If this is true, it puts us physicists under the obligation to explain our ideas in a not-too-technical language."

When Born was writing in the early part of this century, many of his colleagues were also trying to untangle and explain the philosophical implications of their own scientific revolution—the one that rode on the wave of relativity and quantum theory and completely changed the way we think about everything from time and space to energy and matter. J. Robert Oppenheimer wrote this in his book *Science and the Common Understanding:* "The discoveries of science, the new rooms in this great house, have changed the way people think of things outside its walls. . . . It is my thesis that [these discoveries] do provide us with valid and relevant and greatly needed analogies to human problems lying outside the present domain of science or its present borderlands."

Probably the best popular book on Einstein's theories, *The Universe and Dr. Einstein* by Lincoln Barnett, was published in 1948 when Einstein was still alive and the ramifications of the "new physics" were still being unraveled. (Of course they are still being unraveled, but there seemed to be more of a sense then that these subjects were somewhat wet behind the ears, and needed exploring and explaining.) In his introduction, Barnett explained his reasons for writing:

> Today, most newspaper readers know vaguely that Einstein had something to do with the atomic bomb; beyond that his name is simply a synonym for the abstruse. . . . Many a college graduate still thinks of Einstein as a kind of mathematical surrealist rather than as the discoverer of certain cosmic laws of immense importance in our slow struggle to understand physical reality. He may not realize that Relativity, over and above its scientific import, comprises a major philosophical system which augments and illumines the reflections of the great epistemologists— Locke, Berkeley, and Hume.

The idea that science is inseparable from philosophy is a theme that pervades this book. Yet I have to admit another prejudice that colors almost everything I write about science—a preference for what Victor Weisskopf often calls the "old stuff" (somewhat jokingly, I think, since some of these ideas are barely fifty years old, and the implications of quantum theory and relativity have hardly been integrated into the popular culture).

In short, this book is not about black holes. Or quarks, or antimatter, or superconducting magnets and space shuttles. It touches on these things and they are interesting enough in their own right. But concentrating on them tends to promote the feeling that science is something outside our everyday experience. As a friend likes to point out, people talk about traveling into "outer space" as if it were some strange and exotic landscape. And yet *we are living in outer space all the time.* We are spinning around on a big rocky ball, hanging upside down (if you can tell me which direction "down" is), held on by a force which nobody quite understands, and protected and nourished only by the thin seal of the sky. Science is no more "inaccessible" than looking out the window and wondering why a tree branches in a certain way or why (to ask an old but still wise question) the sky is blue.*

I like old stuff because it takes no more knowledge or equipment than that accessible to, say, someone taking a flight on a commercial airliner. Three or four times a year, I find myself a passenger on one of these "space" trips, forty thousand feet in the air and awed, as always, by the oddity of jet-aged flight: my five-hundred-thousand-pound airplane held aloft by unseen forces, no strings attached. The earth below no longer a world, but a spinning sapphire floating in space, veiled by an improbable icing of clouds. From this perspective, you can clearly see its curvature, be amazed by its smallness—a soft, sunlit spot of blue in a great dark void.

At such an altitude, I can look up and see that I am practically bumping my head on the ceiling of space; another forty thousand feet or so more and already I would be in the daytime darkness—a rocketwoman. I look down into a bubbling cauldron of clouds, pierced by an occasional unfriendly mountain peak. We seem to be skimming the surface of an angry sea, and I think it unbelievable that delicate life forms can exist down there—much less build great, noisy aluminum grasshoppers like the one I am flying in to pick them up and hop them from one port to another. Once again, I am reminded that science is stranger than fiction.

On a recent trip, I traveled with my six-year-old son. While he watched the movie, I oohed at the great white splotches of salt on the surface below, like so much spilt milk—the salt of the earth. I

---

*Sometime ago, my young son came up with an interesting variation of this: What color is air? After much thought, we finally came up with the answer: blue. Air is blue for the same reason that the sky (which is made of air) is blue—namely that clusters of air molecules scatter blue light more than all the other colors in sunlight. Air doesn't look as blue as the sky only because there isn't as much of it in a small space. (You might say that air is very, very, very, very light blue.)

aahed at the jagged white spine of the Rockies that slashed the continent down its middle, the twisted river veins that brought life to the blue-blooded planet, the valleys etched so deeply you could almost feel the pain it took to carve them out of the living planet's face. Unable to contain myself, I urged him to look down. He seemed not at all surprised, only vaguely interested. Finally I asked him what he thought was holding the great plane and its hundreds of people up in the sky. As if he were talking to a child himself, he answered, "Air, of course."

Of course.

At that moment, the stewardess came by and asked me to lower my window shade. Other people, she said, wanted to watch the movie. Poor old earth, I thought. Pockmarked by comets, wrinkled and worn by wind and rain, patches of new green growth poking up everywhere, incredibly, in the ashes of the old—and no one to admire your dignified beauty. Talk about taking things for granted!

Thoreau knew that nature was a "wizard," but we seem to have forgotten. "The evolution of a lifeless planet eventually culminates in green leaves," writes naturalist Loren Eiseley. "The altered and oxygenated air hanging above the continents presently invites the rise of animal apparitions compounded of formerly inert clay. Only after long observation does the sophisticated eye succeed in labeling these events as natural rather than miraculous."

Or as physicist B. K. Ridley puts it:

If we were not so used to things dropping when we let go of them, it would seem quite magical that motion appeared out of nothing. That bodies acquire motion from a palpable push is utterly familiar, with a familiarity that stretches back to infanthood. But things in a force field start to move without any visible pushing on them. . . . We can describe what happens quite accurately and we think we understand. But, really, we do not. The invisible influences of gravitation and electromagnetic fields remain magic; describable, but nevertheless implacable, nonhuman, alien, magic.

Well-meaning airline stewardesses aren't the only ones who conspire to prematurely shut our windows on the world. An artist friend of mine, Bob Miller of San Francisco, once made a canvas of reflected light out of a collection of Christmas tree balls arranged in a rectangular box. The silvery balls reflect the images of people looking at them into infinity, crinkle them up at the edges until they disappear. A few years ago, he stuck his box of reflections on a long "stem," planted it in a flower box, surrounded the bottom with junipers, and displayed it at a local art fair. A small child became enchanted with it and asked his mother, "What's that?" Bob happened to be standing nearby when she answered, yanking the child away, "I don't know, but it has something to do with physics."

Bob Miller's Christmas Tree Balls: "It has something to do with physics."

Not that I blame her. Anyone who has sat through a typical physics class and suffered through pulleys and inclined planes has a right to be turned off for life. (Unless they were lucky enough to see Don Herbert as Mr. Wizard on TV during the 1950s and 1960s as I was!) But writing off science can have serious consequences. Most front-page stories today have something to do with science. Our lives are controlled by technology. Yet we do not understand acid rain or how airplanes fly or the basics of electricity. And the

more we view science as a subject fit only for weirdos and geniuses, the less we feel inclined to do anything about our ignorance.

(It's true, of course, that no one can or should be interested in
everything, much less try to keep up with it. I make no effort to
keep up with the latest in mathematics or popular music or professional football—to name a few. But not too long ago I found myself
mired unnecessarily in a monumental traffic jam because I didn't
know it was Super Bowl Sunday.)

Willy-nilly, even the most esoteric aspects of science affect us
all, almost all the time. Gravity, electricity, and even nuclear
forces work on us constantly. Even such elementary constituents
of atoms as quarks and the rarely seen and elusive W particles
have more to do with our daily lives than we usually think. As
physicist B. K. Ridley writes in *Time, Space and Things:* "One
cannot deny a sense of unreality when one contemplates such a list
of subnuclear things. What possible interest can they evoke? . . .
Why bother with [them] when they are so far removed from everyday things?"

Then he answers himself: "The fact is that they *are* everyday
things. As you read this, particles of unimaginable energy are
pouring into the earth's atmosphere and knocking nuclei to bits.
Some of those bits are going right through you or bashing into one
of your own private nuclei at this moment." This nuclei bashing, he
goes on to point out, is responsible for the sunlight we bask in, the
source of all our energy. It is also integral to the evolution of life,
and the original formation of all the elements.

In the end, however, there's an even more fundamental reason
for my concentration on everyday "old" stuff: What's fascinating
about science at the forefront is only an embellishment of what's
fascinating about science in everyday life. Gravity is at the bottom
of black holes. In fact, a black hole is just a way of looking at what
gravity would do if pushed (or pulled) to extremes. Black holes
may or may not exist, but gravity itself is a deep (and unsolved)
puzzle. How does it get from there to here? What is its source?
Where did it come from? Is it truly the geometry of the space we
live in?

In the same way, the behavior of supercold materials seems
sometimes supernatural: the so-called superfluids that exist only

around absolute zero (minus 459 degrees Fahrenheit) can flow up
and out of bottles, down through the bottoms of ceramic containers; superconducting metals can carry an electric current forever
without offering the slightest resistance. At the hottest end of the
temperature scale, matter also acts strangely. Atoms fall apart,
forming superhot, electrically charged plasmas that ignite fusion
fires like those in the center of the sun. To tame these elusive, too-
hot-to-touch gases, physicists have constructed huge magnetic
bottles, but a plasma is a slippery thing, and hard to contain. Plasmas are the stuff of stars.

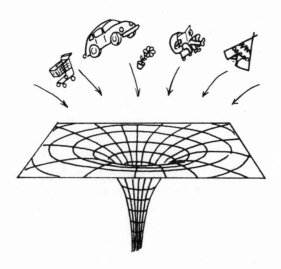

Gravity is at the bottom of black holes.

Yet all these exotic forms of matter are nothing more than extensions of the familiar spectrum of states of matter—from solid to
liquid to gas. And it is impossible to appreciate either supercon-
ductivity or plasmas without understanding the transformation of
water to ice and steam.

Some people say that subjects like gravity or the states of matter are too old and too simple to be interesting. But in truth, no
science is really that simple. "Most of us are in daily contact with at
least as much that we do not understand as were the Greeks or
early Babylonians," my friend the physicist likes to say. "Yet we

have learned not to ask questions about how the power steering on our cars works or how polio vaccine is made or what is involved in the freezing of orange juice. We end up in the paradoxical situation in which one of the effects of science is to dampen curiosity."

If simple science is uninteresting, it may be only because we have been made ashamed of asking all those simple, "obvious" questions—like "Why isn't the emperor wearing any clothes?" We still don't know, after all, where the moon came from, or how life arose on earth, or why the speed of light is 186,000 miles per second. We don't know why people respond to music or whether it will rain tomorrow in Minneapolis. We don't know the nature of evil or of the force that binds quarks together. "The genius of men like Newton and Einstein," writes J. Bronowski in *The Ascent of Man*, "lies in that: they ask transparent, innocent questions which turn out to have catastrophic answers. Einstein was a man who could ask immensely simple questions."

Asking good questions is much more important than finding good answers because the answers are always changing. In fact, one of the main differences between science and philosophy or religion is that scientific beliefs are tentative. By definition, they are incomplete.

This book is not intended to offer definitive answers. But I do hope it raises some interesting questions. (As Mark Twain said, the best thing about science is the enormous amount of conjecture one earns for such a trifling investment of fact.)

Most of all, I hope it clarifies the connections between physics and the other aspects of our lives—because even if these connections are not always "real" in a scientific sense, they are certainly alive and at work in our consciousnesses.

I. THE SENTIMENTAL FRUITS OF SCIENCE

Popular science writers are forever proclaiming the profound importance of such matters as the ultimate fate of the universe, or the events that took place during the first billionth of a second of time. We often write as if people were poised on the edges of their proverbial seats, anxiously waiting to learn whether or not the proton will decay (in 10^{32} years!)* or whether there is mass tucked away mysteriously inside neutrinos. Breathlessly, we keep them up to date on such esoterica as the search for "bottom quarks" and "intermediate vector bosons."† Sometimes I wonder how many readers struggling to get their socks on in the morning doubt whether exploring these far-out corners of the universe is necessarily so important. It is surely not obviously useful.

Sixteen years ago, I sat in the musty lobby of a small hotel in the Soviet industrial city of Kharkov, having just returned from a visit to a collective farm. Thirty or so Americans and Russians squirmed together in the tiny uncomfortable room to watch on a scratchy black-and-white TV as two American astronauts walked on the moon. The image on the screen was a barely discernible blur; yet it was quite clear that both Russians and Americans were deeply impressed by these first tentative extraterrestrial steps. We agreed completely when Neil Armstrong described it as a giant leap for humankind.

It was many years later before I learned that some of my best friends back home had considered the whole venture a waste of time, a squandering of scarce resources. Worse, even supporters of the space program seemed to be countering with all the wrong arguments. Recently I was reminded of this when a man from Grumman started drumming up support for the shuttle by passing around samples of Mylar jogging suits and talking—again—of Tang. Then the editor of a major woman's magazine told me that I should write only of the practical aspects of science: "People want information they can use."

It's true, of course, that science has produced a prodigious array of practical fruits—and those are the fruits that we hear about.

---

*10^{32} is the number 1 followed by thirty-two zeros. To give you an idea of how large that is, consider that a billion is the number 1 followed by only nine zeros. Physicists think the entire universe is only between fifteen and twenty billion years old. Waiting for a proton to decay is like waiting for Godot.

† As far as physicists know, the quark is the most elementary constituent of matter. Quarks make up protons and neutrons, which make up the atomic nucleus. The five known quarks are called up, down, strange, charmed, and top (also called truth). The theory requires a sixth quark (bottom, or beauty) which has not yet been found. For a description of bosons, see "Forces, Motives, and Inertia."

The spin-offs from the space program alone have brought us everything from improved mapping of the earth's surface to a deeper knowledge of what lies beneath its crust; from advanced microprocessors to zero-gravity production lines for making flawless crystals and pure drugs; from space-age medical technology to steamlined sailboats and lighter, more durable sportswear. But any spin-off is, by definition, a somewhat peripheral point. Science is not the same as technology. There's nothing *practical* to be gained (for a long while anyway) from probing protons or searching for signs of extraterrestrial life.

There is, however, a far more central point to science. The point is what my friend the physicist calls "the sentimental fruits of science." "Science is useful not only in a practical way," he says,

> but also in that it determines how we think and feel. Religions have always embodied a view of nature. Even the Bible begins with an account of cosmology. Today, such thoughts about nature come primarily from science. They are as imaginative and as fantastic as ever. But today people ascribe a very limited role to science. They continue to talk of the arts and music as culture, but neglect the fact that our view of ourselves and our perception of what our world is like are equally and vitally a part of culture.

Victor Weisskopf calls modern science the "greatest cultural achievement of our time." Isaac Asimov describes the telescope as an instrument that dramatically changed our *cultural* history (emphasis his): "When Galileo looked at the moon with a telescope and saw mountains, craters and 'seas,' that was the final piece of evidence in favor of a plurality of worlds. Earth was not the only object on which life could conceivably exist." The telescope so expanded our view of the universe that "the great man-centered drama of sin-and-redemption, constructed in earlier times, looked puny against the new universe."

Or as Stephen Jay Gould has noted—in this case in reference to Darwin: "Major ideas have remarkably subtle and far-ranging extensions. The inhabitants of a nuclear world should know this perfectly well, but many scientists have yet to get the message."

Last Christmas I took my son to see the show at the Hayden

Planetarium at the Museum of Natural History in New York. It was an excellent opportunity to enjoy some sentimental fruits of science. As the show opened, we looked up at the spotty cover of stars we see every night, surrounded by the bright lights and tall buildings of the city on the horizon. Suddenly the "city lights" went out, and the starry skylights lit up with a splendor that made the audience gasp. I was immediately overwhelmed with a sense of what it must have been like to live on a flat earth under such a lively canopy of stars, where the constellations were as real and as close as mountains and meadows, rather than on a spinning ball floating at the edge of an ordinary galaxy, one of several hundred billion in the universe.

It also reminded me that seeing stars is all a matter of perception. We rarely wonder where the stars spend their days, but turning on the sun has the same effect as turning on the lights in the city: the canopy is there all the time. But like anything else, we can't see it when so much "extraneous information" gets in the way.

The stars are an obvious place to start searching for "sentimental fruits." After all, it was an understanding of the motions of the planets and stars that first plucked people from their center-stage spot in the solar system. Although some people had obviously figured out as far back as the third century B.C. that the earth orbited around the sun (and not vice versa), this wasn't incorporated into the popular culture until well after Copernicus—in the sixteenth century. Conventional wisdom put the earth at the center of things; the universe was here for us. Imagine what that meant for people's sense of destiny, personal responsibility, and awe.

In some ways, however, the things we've learned through science have shown us that the earth is more central than ever—that life is even more precious because of its improbability. Our planet is but a fragment of an exploding star, basking in the light of a second-generation sun; its composition a consequence of a slight contamination by foreign elements in the original hydrogen gas cloud that formed the solar system. In a very real sense, our rocky home is the sediment that sank to the bottom (really the center, but then "down" is always toward the center of the earth) when the lighter elements were blown or boiled away. Animals arose on land only after plants accidentally polluted their environment with a "poison" (to them) called oxygen. This knowledge has not necessarily made us more or less humble, but it has certainly, says my friend the physicist, "changed the nature of our humility."

A closer look at the cosmos also shows that the seemingly static universe is dizzy with change; even the stars use up their resources, die, and are born again. But the sometimes violent births and deaths of stars, the continual evolution of the universe, was unknown only a few centuries ago. Until the time of Galileo it was simply assumed that the stars we see today are the same stars that existed at the beginning of time, the same stars that would exist forever. The universe—like the status of slaves—was fixed.

At the opposite end of the size scale, the twentieth-century development of quantum theory shattered the notion that atoms behave like billiard balls, and that everything they (and therefore we) do is predetermined. A great deal of uncertainty lies at the heart of atoms; the meaning of a seemingly simple idea like cause and effect turns out to be immensely rich and complicated. But the result is that things today look far more flexible than they did in Newton's clockwork, preset universe.

Charles Darwin's formidable (and in some corners still forbidden) fruit was the knowledge that species, like stars, can change. The forms of life that inhabit the earth are not immutable. We, like the universe, *evolve*. Strict biblical creationists wouldn't be creating such a stir about Darwin if the question of where we came from and what our ancestors looked like wasn't an issue that itself stirred deeply within our souls.

It wasn't until the last century that Darwin for the first time firmly established the scientific basis for a kinship among all living things—a vastly extended family tree. And if Darwin's ideas still remain unacceptable to certain people, it is not because something was wrong with his science. It is rather, says Gould, because of the "radical *philosophical* content" of his theory: "The true Darwinian spirit might salvage our depleted world by denying a favorite theme of Western arrogance—that we are meant to have control and dominion over the earth and its life because we are the loftiest product of a preordained process."

Darwin's theories have been widely misinterpreted to mean that people are the top dogs on the evolutionary tree. But his true point was just the opposite—our intimate connection with all other forms of life. Gould has devoted many an essay to discrediting those who misuse Darwin in order to justify the dominion of whites over blacks, rich over poor, men over women. All Darwin said was that through evolution all forms of life become well adapted to their particular environment. There was no mention of "higher" or "lower" species.

Naturalist Loren Eisely puts this quite beautifully:

> We today know the result of Darwin's endeavors—the knitting together of the vast web of life until it is seen like the legendary tree of Igdrasil, reaching endlessly up through the dead geological strata with living and related branches still glowing in the sun. Bird is no longer bird but can be made to leap magically backward into reptile; man is hidden in the lemur, lemur in the tree shrew, tree shrew in reptile; reptile is finally precipitated into fish.

Understanding evolution does not lead to the inevitable conclusion that people are but well-bred apes. Rather, it makes us appreciate the care that a million years of adapting to nature has put into fashioning every creature on this planet—including every man, woman, and child. Even more, it makes it impossible to deny the intimate connections between the members of our human species. Young, old, male, female, black, white, Israeli, or Syrian, we are all incredibly similar. Racism and religious wars are both waged under the banner that "other" people are different. Yet "this new insight of science," my friend the physicist says, "has made it much harder to believe that other people are really, fundamentally, different from ourselves."

Even our view of strictly physical forces such as gravity can have a profound effect on the way we view ourselves. Newton blew up a storm in the prevailing cultural winds of the eighteenth century with his Universal Theory of Gravitation not because he "discovered" gravity (everyone knew things fell toward the earth) but because he discovered that gravity was universal. Before him, it was assumed that the laws of nature on earth were fundamentally different from those in the heavens. Newton showed that the fall of the apple and the orbit of the moon were controlled by the *same* forces.

In this sense, the need to go to the moon or smash atoms is on a par with the need to have natural history museums: science provides a handle on who we are and how we fit into the scheme of things. Materialistically speaking, the Apollo landings certainly were a waste. Well-equipped robots could easily have accomplished the same things. But the moon has simply not looked the same since we can look up and say, "Someone was walking around up there." The idea is *emotionally* entrancing—like E.T. The moon shots were *sentimental* journeys. They were important not because they were technologically fruitful, but because they were *awe*-inspiring.

It's the same sense of awe that comes from stargazing or from lying on your back and feeling that you might at any moment fall "down" into outer space, that you are stuck to this earth only by the invisible glue of gravity. Or the feeling that you get when you examine your finger and imagine the unseen activity that goes on inside a single cell—Lilliputian worlds within worlds. It makes you stop and think: to realize that even the hardest rock is almost entirely empty space—a lattice of elusive bundles of energy held together by bits of electric charge; to know that the sky is only as high as the comparative thickness of your condensed breath on a marble. What a thin skin our planet has!

I have a poster on my office wall that shows a spiral galaxy with a large arrow pointing to a middle-sized star on one of the outer rims; words attached to the arrow read, "You Are Here." The small print on the bottom tells me that there are some hundred billion such galaxies in the universe, each containing at least a hundred billion stars.

This kind of information should not leave us feeling that people and their achievements are insignificant. Weisskopf titled his popular book *Knowledge and Wonder* because wonder is every bit as important as knowledge in the attempt to understand nature. "A good society is one which celebrates its own existence," he says. "It's important not to leave all the awe to the religionists." When his students get depressed, he tells them there are two things worth living for: Beethoven and quantum mechanics. What he means is that the beautiful things created by nature and by people are worth our constant respect and admiration (not to mention preservation).

Contemplating our cosmic navels is a good thing if only because it feeds our souls—and a lot better than Tang. The sight of Venus dangling near the crescent moon on a clear night like a diamond earring stirs our emotions in the same way great art does. Recently I sat in box at Lincoln Center and listened to Joan Sutherland sing Lucia, amazed at the emotional impact produced by the subtle stirring of sound waves inside my ear. It was not unlike the feelings people get from looking out over the ocean, or listening to a bird sing, or probing the innards of protons.

I have a friend who likes to ask the following "science" question: How would you hold one hundred tons of water in thin air with no visible means of support? Answer: Build a cloud. What he's saying is no more elaborate than "Gee whiz!"

Understanding our place in the sun requires an understanding of the sun's place in the solar system, the cycles of the sky, the nature of the elements, and the improbabilities of life. If what we learn leaves us a little stunned by our limitations and potentials, so be it. Science gives us a sense of scale and a sense of limits; an appreciation for perspective and a tolerance for ambiguity.

(I'm not saying that science isn't a part of our everyday lives in the ordinary sense, either. If it weren't for the technology that has come from science, I wouldn't be typing on this typewriter;* I would be scrubbing diapers on a washboard instead of covering my baby's bottom with layers of disposable plastic. I feel a little bit guilty about the plastic, of course. I know it pollutes the environment. Perhaps in talking about the fruits of science, we should have three categories: practical fruits, sentimental fruits, and rotten fruits. The rotten ones have the increasingly scary potential to spoil all the rest. But it's undoubtedly easier to fight back against their dangers with knowledge rather than ignorance. In any event, the understanding of the nucleus doesn't build the nuclear bomb.)

Nobel laureate Steven Weinberg concluded his account of the First Three Minutes in the life of the universe with a curious statement: "The effort to understand the universe," he wrote, "is one of the very few things that lifts human life a little above the level of farce and gives it some of the grace of tragedy."

I prefer a more positive version of the same sentiment that

---

* Actually, the typing of this manuscript finally compelled me to buy a computer.

came from Robert Wilson—the sculptor and physicist who built the giant atom smasher at the Fermi National Accelerator Laboratory near Chicago. He expressed it in response to the continual questions of a senator who demanded to know what probing protons had to do with the national defense:

"Is there anything connected in the hopes of this accelerator that in any way involves the security of the country?" asked the senator.

"No, sir, I do not believe so," responded Dr. Wilson.

"It has no value in that respect?" asked the senator again.

"It only has to do with the respect with which we regard one another, the dignity of people, our love of culture. It has to do with these things: Are we good painters, good sculptors, great poets? I mean all the things that we really venerate and honor in our country and are patriotic about.

"In that sense, this new knowledge has all to do with honor and country but it has nothing to do directly with defending our country—except to help make it worth defending."

2. SEEING THINGS

Several years ago, when I first began delving into the curious Alice in Wonderland world of particle physics—that subatomic never-never land inhabited by "quarks" and "gluons,"* entities "strange" and "charmed"—I asked my friend the physicist how anyone could believe in such seemingly ephemeral objects, things that no one could ever really see. And he answered: "It all depends on what you mean by seeing."

*A gluon is a particle that carries the so-called "strong" force that holds quarks together to form protons and neutrons. See "Forces, Motives, and Inertia."

Like many people, I always feel somewhat skeptical when I hear physicists confidently claiming to have "seen" particles effervescing into existence for a mere billionth of a second, or massive "quasars"* teetering twelve billion light-years away at the very brink of space and time. I know, for a fact, that they have "seen" no such thing. Quarks and quasars are invisible to the naked eye— even to the finest instruments. At best, the physicist has seen a bump on a curve plotting the ratio of various kinds of particles produced in a nuclear collision; more often, such a "sighting" is in truth a conclusion laboriously drawn from long hours of computer calculation and long chains of inferences and assumptions. Hardly the sort of thing to inspire an exultant "Eureka!" (Or even "Land ho!") Sometimes the things that scientists say they see are so removed from actual quarks or quasars that one wonders if they (or we) should believe their eyes.

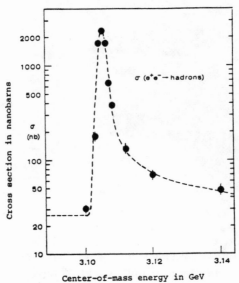

The J/psi particle as seen by physicists at the Stanford Linear Accelerator Center. It was not called a "particle," but rather the "ψ/J (3095) resonance." See "Sympathetic Vibrations," Chapter 12.

*Quasar is short for quasi-stellar object. Quasars are sources of immense outpourings of energy found far away from us in space and time. Their exact nature is still unknown.

Physicists "see" exotic particles by bombarding them with other particles and analyzing the patterns created as the particles bounce back into their electronic detectors. The first person to "see" the atomic nucleus used much the same method—except that the electronic detector used was the human eye. During World War I, Lord Ernest Rutherford aimed a beam of particles streaming from a radioactive rock toward a thin sheet of gold foil. Most of

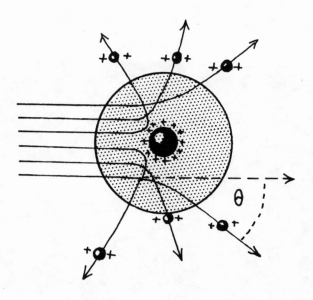

How Rutherford "saw" the nucleus.

the particles passed right through. But some—surprisingly— were scattered through very large angles and a few were even reflected *backwards*. "It was about as credible," said Rutherford, "as if you had fired a fifteen-inch shell at a piece of tissue paper and it came back and hit you." From this Rutherford concluded that the atom was not a uniform mass, as previously thought, but something rather more like a miniature solar system—with almost all of the mass concentrated in a small, central nuclear "sun." Most of the particles passed right through because there was very little inside the gold atoms for them to hit; if a particle did chance to hit the nucleus, however, it could be reflected backwards like a ball hitting a brick wall.

Today, physicists similarly "see" all kinds of particles by bombarding all kinds of targets with all kinds of other particles—and they use sophisticated electronic equipment to analyze their results. But in truth, seeing inside an atom is not so different from seeing a friend or a building. It only *seems* unreal and abstract because the details of the process are slightly less familiar: The way we go about seeing things in everyday life is scarcely what you'd call "direct."

Physicists "see" exotic particles by smashing together other kinds of particles and analyzing the debris. The SPEAR ring at the Stanford Linear Accelerator Center (above) "discovered" many exciting particles in its day, but it's almost obsolete compared to some the new "colliders" that whirl particles around in seventeen-mile-long rings before smashing them together.

I see my typewriter, for example, because rapidly vibrating atoms in the filament of the light bulb overhead and in the sun outside my window some ninety-three million miles away are sending out streams of light particles called photons, some of which are showering down right now on the typewriter's surface. A few of these photons collide (as in an atom smasher) with the molecules in the painted metal and the plastic-coated keys, and bounce back in the direction of my eyes. If they penetrate the pupil, they will be focused by a lens onto a light-sensitive screen (the retina)—a sophisticated electronic detector that passes along information about the photons' energies, trajectories, and frequencies to my brain in

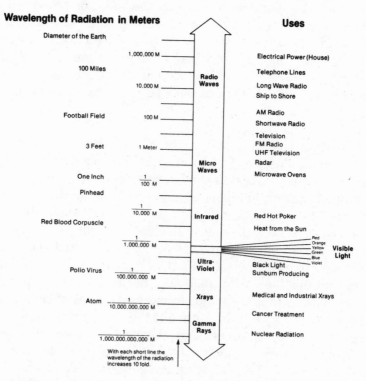

Visible light is but a small sliver of the electromagnetic spectrum.

the form of digital bits. On the basis of laborious calculations and long chains of inferences and assumptions, my brain concludes that the light patterns represent something hard and cool and heavy and solid that translates passing thoughts into printed words.

Of course, my eyes, like particle detectors and telescopes, are tuned in only to the narrowest band of information coming from the outside world. The pupil is but a tiny porthole in a sea of radiation. In a universe alight with images, we are mostly in the dark. Human eyes respond only to those electromagnetic vibrations that make waves between .00007 and .00004 of a centimeter long. Yet, as I type, I am bombarded by other kinds of light waves as small as atoms and as large as mountains, coming from the far reaches of space, from the inside of my own body, from the TV transmitter twenty miles away. I know that these signals are there, in the room with me, because if I flip on the radio or television I will suddenly be able to see or hear them—in the same way that visions suddenly "appear" before me the minute I open my eyes. If I had still other kinds of detectors (I can sense some of the infrared radiation on the surface of my skin as heat), I could pick up still other kinds of signals. Yet we walk through this dense web of radiant information every day without being the least aware of its existence.

Radiation is only one kind of information to which we are mostly blind. We are deaf to most of the sound around us. Our chemical senses (taste and smell) are extremely limited compared to those of a plant, or a cell, or a dog. We can barely perceive the difference between hot and cold, to the extent that a blindfolded person can't tell whether he or she has been burned on a hot iron, or on dry ice. Even our perception of forces is curtailed by our size. While we can easily sense the pull of gravity, we are almost completely insensitive to the pulls and pushes of air resistance and surface tension that are major forces in the lives of cells and flies. We don't have to push the air out of the way to walk through it as the gnat does; on the other hand, the electrical force of cohesion is relatively so much stronger for a small creature that the fly can crawl up the wall, completely ignoring gravity. To say we are narrow-minded (or at least "narrow-sensed") is the least of it.

Naturalist Loren Eiseley writes of coming upon a spider web in a forest. The spider web is confined to its own two-dimensional universe, totally oblivious to the plants or people around it, even to the pencil Eiseley uses to reach out and pluck it: "Spider thoughts in a spider universe—sensitive to raindrop and moth flutter, nothing beyond, nothing allowed for the unexpected, the inserted pencil from the outside world."

We live in a spider web too—a three-dimensional spider web spun in the unseen context of our four-dimensional space—and we are only beginning to become aware of the vast universe outside. Our perceptions of time and space are largely limited to things in our own experience, of our own relative size. We find it almost impossibly difficult to understand numbers much larger than those we can count on our own fingers and toes, or spans of time much longer than our lifetimes. Looking outside our spider web takes an enormous leap of the imagination. From our spider point of view, the world is clearly flat. It is also obviously motionless. It was probably Galileo who first proposed the idea that in fact we could not tell by experiment alone whether we were moving or not. Put yourself inside the closed cabin of a steadily moving ship, he said. Allow small winged creatures like gnats to fly around, watch fish swimming in a bowl, toss objects back and forth, notice how things fall to the ground. No matter how many experiments you perform, "You shall not be able to discern the least alteration in all the fore-named effects, nor can you gather by any of them whether the ship moves or stands still."

What you perceive as "standing still" at the equator is in reality a rapid spin around the earth's axis at the dizzying rate of 1,000 miles per hour. In addition, the entire spinning earth is whizzing around the sun with a speed of almost 20 miles per second. The solar system itself is moving with respect to the center of our galaxy at 150 miles per second, and our galaxy—the Milky Way—is rushing toward our neighboring Andromeda galaxy (from its point of view) at 50 miles per second. And that's not all: if you looked at the earth from a far-off quasar, you might see us speeding away at 165,000 miles per second, close to the speed of light.

What Galileo stumbled upon in stepping out of his spider web was what Einstein would later refine into the theory of relativity. Einstein saw that there were other things we couldn't perceive, like the elastic nature of space and time. We were being deceived by our senses into thinking that our own Euclidean three-dimensional geometry was the geometry of the universe. Einstein was out of his senses. He saw that it didn't matter whether or not our own crude perceptual instruments could pick up the tiny increases in the mass of objects as they moved faster.* It was a limitation of our clumsy instruments. Today's modern particle accelerators routinely push particles to speeds almost as fast as the speed of light—and see them gain forty thousand times their initial weight in the process!

One of the most important things we seem not to perceive is the process of exponential growth.† We think we understand how compound interest works, but we can't get it through our heads (literally) that a fine peice of tissue paper folded upon itself fifty times (if that were possible) would be seventeen million miles thick. If we could, we could "see" why our income buys only half what it did four years ago, why the price of everything has doubled in the same time, why populations and pollution and nuclear bombs seem to (in fact, really do) proliferate out of control.

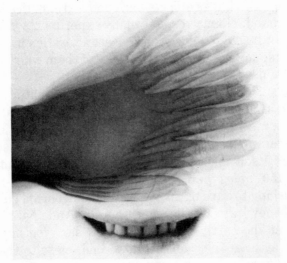

A simulation of the Cheshire Cat effect; a real photograph would be impossible to take because the image that you see is constructed in your brain—a product of the mind's eye.

In addition to all the things that we *can't* see are the things that we *don't* see because we choose to ignore them. Right now, I am choosing to ignore the sounds of my own breathing, the touch of

*See "Relatively Speaking."
† See "Small Differences."

the wedding ring on my finger, the sight of the glasses that are right in front of my nose—and even the nose on my face. While I focus on at most a word or two at a time, I am deliberately blurring all thoughts of the baby downstairs, the situation in El Salvador, the impending crisis in the economy, the leak in the sink, and the friend who is coming for a visit tomorrow night. Shutters and pupils are meant not to let information in, but to "shut" it out. As anyone knows who has ever held a camera, too much information is easily as blinding as too little. If you played all nine Beethoven symphonies at once, you would hear nothing but noise.

But deciding what to turn on and off and when is a dangerous business—especially since mostly we are unaware of it. Recently, four people sat in my living room directly underneath a loud antique clock. At 3:05 I asked whether the clock had struck three. Two people insisted it had, and two insisted it hadn't.

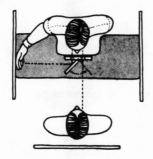

A bird's-eye view of the Cheshire Cat exhibit. Mirrors are arranged so that one eye of a viewer sees his friend in a direct line of sight; the other eye sees light reflected in a mirror from a white wall. If a hand moves in front of the wall, the brain temporarily erases the image of the friend.

Our eyes automatically erase the oversized feet and errant fingertips and out-of-focus images that are clear to the more objective camera's eye. Distractions erase reams of information—which is one reason there is a difference between "listening" and "hearing." It is a physical fact that you cannot listen to even two conversations at once, or see more than the narrowest sliver of a visual field.

There is an exhibit in The Exploratorium, in San Francisco,

called the Cheshire Cat, in which, through an arrangement of mirrors, you sit so that one eye sees a face while the other eye looks at a blank wall. Your brain "sees" the face, erasing the "uninteresting" wall. But if someone makes a motion in front of the wall, your brain shifts its focus of interest away from the face and to the wall. It (you) becomes distracted. Now you have to *choose* which image to see. The brains of most people blithely erase parts of the face—leaving sometimes the mere glimmer of a smile (hence the exhibit's name). You can keep the face in view—or at least parts of it—if you try very, very hard. But focusing on anything amidst all that distraction becomes amazingly difficult.

Vases or faces? The answer depends on what you perceive as the background—the black spaces or the white. Photographer Zeke Berman created the illusion by using silhouettes of real people; his own silhouette appears at bottom center.

Sorting information from "noise" is one of the most important processes in all of perception. Yet it is also obviously a minefield of potential mistakes. There is a simple and striking illusion in which two facing profiles suddenly appear as a vase, which just as suddenly can fade again into two facing profiles. You cannot see both vase and faces at once—because you cannot see something as "background" and "foreground" (or information and "noise") at once. Whatever is in the background becomes truly as invisible as if it weren't there—even though you may be looking right at it—just as people in our lives really "vanish" as they fade into the background of our consciousnesses.

Mostly what disappears are the sights we get used to—like our noses and eyeglasses, but also our sunsets and even the sounds of our children. Steady signals fatigue our senses, numbing our powers of response. Dogs will sleep through all kinds of everyday noises only to snap into alertness at the soft step of an intruder; parents have been known to snore through sirens and garbage trucks only to awake alert at the merest whimper from their newborn baby.

These kinds of sensual atrophy are largely learned. But others are completely automatic. That is, some kinds of signals actually fatigue our physical sensors to the point that we are no longer able to perceive them. Perhaps the most common example of this is the everyday afterimage. If you stare at a bright light or open your eyes to the streaks of sunlight coming through the edges of the shades first thing in the morning, you are likely to look away only to see the image lingering in your field of vision. The afterimage is dark where the original image was bright—because it corresponds to those places on your retina where the sensors have been bleached by the light. For some moments, they can no longer respond. They can no longer send the signal to your brain saying, "white wall here" or "blue sky here." The rest of your retina responds normally, so what you see is a normal background with a dark image of the original bright "flash" imposed on it.

Some painters actually color their work with afterimages in mind. For when your eyes fatigue of one color, they will see its complement. Say, for example, you stare for fifteen seconds or so at a bright-red area on a painting or on a wall. The red sensors in your eye fatigue. If you then look at a white wall, your eyes will send the following message to your brain: white *minus* red. Since white minus red is green, green is the color you see. (If you stare at a green spot, you see a red spot when you look at the white wall. And so on.)

Your motion sensors work much the same way. If you spin around the room in one direction, your motion sensors soon lose their ability to respond to the steady signals: they no longer send a message to the brain that you are turning clockwise; turning clock-

wise has become synonymous with "stopped." When you stop spinning, the fatigued sensors respond by sending a message to the
brain that says: no longer stopped, spinning in opposite direction.
You sense yourself spinning *counterclockwise*. If you stare at falling water for fifteen seconds or so and then switch your gaze to the
ground, it can make the ground seem to "fall up," a phenomenon
appropriately known as the waterfall effect.

Sensory fatigue, that is, can sometimes cause you to perceive
the *opposite* of the signals you actually receive. One wonders if this
has some bearing on the way the important social issues of one
decade seem to fatigue in the next, producing the well-known
"swinging pendulum" response: how the broad social consciousness
of the sixties reemerged as the narrow self-interest of the seven-
ties, for example. Just as your color sensors fatigue of seeing red,
so your social sensors fatigue of marching for peace, taking care of
poor children, talking of black rights or women's rights. One wonders what (so fatigued) people are really "seeing"—what information is reaching, and registering, in their brains.

People who are brilliant scientists (or writers or parents or doctors or carpenters) are those who have a special talent for keeping
the important things in focus—both separating the signal from the
noise and also knowing when what sounds like noise might contain
the quiet whisper of important information. In a very real sense,
this was the remarkable aspect of the Nobel Prize-winning discovery in 1964 by Robert Wilson and Arno Penzias of the faint radiation that pervades the universe, and is now thought to be a
probable remnant of the primal Big Bang. The two astronomers
tried everything they could think of to get rid of a persistent buzz
in their Bell Labs radio telescope. When they couldn't, something
must have told them that it wasn't just noise.

(Of course, one person's noise is another person's information.
Some physicists send probes high into the upper atmosphere to
catch and analyze cosmic rays. Other physicists interested in
tracking other kinds of particles program their computers to eliminate the annoying cosmic ray "noise.")

The instruments of science have vastly extended our senses.
Indeed, British physicist David Bohm concludes that "science is
mainly a way of extending our perceptual contact with the world,"

its purpose being to foster "an awareness and understanding of an ever growing segment of the world with which we are in contact." Technology has unveiled vast new vistas, opening up untapped realms of time, space, and temperature. To modern telescopes and particle accelerators, the radio waves and gamma rays invisible to us are rich with images. The number of so-called elementary particles has proliferated wildly because the instrumentation to "see" them has gotten better and better. The same is true of the number of stars in the sky, and such strange newcomers to the galactic zoo as pulsars and quasars. We can see out into space, back into time, inside our own genetic structure. We can see what the stars are made of and how a virus looks. We can measure things smaller and larger, colder and hotter, faster and slower than could ever be "seen" before.

This track of a charmed particle was seen with the aid of a very sophisticated bubble chamber.

"How rich we are," writes Guy Murchie, "that we can look on these worlds with the perspective of modern science . . . that we do not have to wonder as did former men whether stars are jewels dangling from celestial drapery or peepholes in the astral skin of creation!"

Our view of the universe is changing so rapidly partly because our ability to see is growing so rapidly. "Early descriptions of the universe are egocentric and based on the physical size and capabilities of man," writes Richard Gregory in his marvelous book on perception, *The Intelligent Eye*. Pre-science philosophy was based solely on human perception. But now we know that there is a lot going on that we can't see *except* through science. "The simple fact that stars exist invisible to the unaided eye made it unlikely that the heavens are but a backcloth for the state of human drama."

Scientific perception has a different authority from personal perception because it can more easily be shared. It's a way of seeing that many people can agree on, or at least, agree on a way of thinking about. But the process is essentially the same: scientists "see" by gathering data, measuring, making assumptions, and drawing conclusions. "Elementary particles don't seem real to ordinary people because they aren't perceived in an ordinary way," says M.I.T. physicist Vera Kistiakowsky. "Something like astronomy *seems* more real because you can see the stars with your own eyes. But even that is mostly inferred. All science involves the interpretation of secondary information."

All *perception* involves the interpretation of secondary information. We are always seeing a great deal more than meets the eye. The light patterns that form on the tiny screens within our eyes are upside down, full of holes and splotches, badly bent out of shape. Most of what we see is in our heads. If I believed my eyes, I would see people shrink to Thumbelina size as they walked away. In fact, all my visions would remain inside my body. For it is our brains that perform the incredible feat of projecting what we see "out there" from the backs of our eyes to some arbitrary place in space. Not only vision, but *all* sensory experience takes place within our bodies. Yet we attribute these properties to objects that exist outside us. We say that ice cream tastes sweet, or the table feels hard, when in fact it is *we* who taste and feel.

I am continually amazed that we are able to extract anything at all from the meager and subtle messages that reach us from the outside world, much less such a rich feast of sensations. Tiny changes in air pressure on my eardrums tell me that my child is laughing, that a friend is calling on the phone, that my flute is out of tune, or that the wind is rustling in the trees—whole symphonies of sound. Sensations such as "soft," "green," or "melodious" arrive at the brain as electrical signals. Taste is the detection of molecular structure on the tip of the tongue; warmth, a calculated difference in the rate of chemical reactions. Galileo recognized that qualities such as color and smell "can no more be ascribed to the external objects than can the tickling or the pain caused sometimes by touching such objects." Democritus wrote twenty-three centuries ago that "sweet and bitter, cold and warm as well as all the colors, all these things exist but in opinion and not in reality." Why do we call white a *lack* of color instead of the brilliant combination of hues it is? Probably because sunlight is whitish, suggests George Gamow in his *Biography of Physics,* and therefore we dismiss the everyday color we are accustomed to as "ordinary."

"We each live our mental life in a prison-house from which there is no escape," writes British physicist Sir James Jeans. "It is our body; and its only communication with the outer world is through our sense organs—eyes, ears, etc. These form windows through which we can look out on to the outer world and acquire knowledge of it."

But that knowledge, Jeans points out, does not necessarily correspond to the kind of information we receive. Ice cream does not taste electrical, even though electrical impulses are all that arrive in our brains. "In any case," said Jeans,

> there is no compelling reason why phenomena—the mental visions that a mind constructs out of electric currents in a brain—should resemble the objects that produced these currents in the first instance. If I touch a live wire, I may see stars, but the stars I see will not in the least resemble the dynamo which produced the current in the wire I touched. . . . We may watch the sparks fly as the blacksmith hammers a piece of iron into a horseshoe, but we must not infer that the piece of iron is an accumulation of sparks, each having the properties of those we see flying through the air.

The raw data need to be translated into a context we can understand. Even language itself is built mostly of "meaningless" words, which can be heard and interpreted according to the way we tune our perceptual antenna. A single word can have dozens of different

meanings and a single collection of sounds can communicate dozens
of different words. William Safire recounts a delightful example of
this in his book *On Language*. Schoolchildren have interpreted the
sounds we take to mean "I pledge allegiance to the flag" as everything from "I pledge a legion to the flag" to "I led the pigeons to the
flag." Children, writes Safire, "make sounds fit the sense inside
their own heads." So do we all.

*The moirés created when two fine curtains overlap are fash-
ioned from unseen threads in the fabric that combine to produce the larger, visible, pattern.*

Our newfound scientific senses are even farther from direct interpretation. The images of quasars seen by telescopes using Very
Long Baseline Interferometry, for example, are really composite
patterns resolved from information recorded separately at individual telescopes as much as six thousand miles apart, synchronized
by atomic clocks and pieced together by computers. They are not
"images" in the ordinary sense, but rather interference patterns
like the rapidly moving moirés you see when two fences or fire-

place grates or curtains overlap—secondary patterns emerging from the combination of two (unseen) patterns.*

This process of abstraction means that the closer we get to reality, the farther removed it seems. "The evolving picture becomes ever more remote from experience," writes Lincoln Barnett in *The Universe and Dr. Einstein,*

> far stranger indeed and less recognizable than the bone structure behind a familiar face. For where the geometry of a skull predestines the outlines of the tissue it supports, there is no likeness between the image of a tree transcribed by our senses and that propounded by wave mechanics, or between a glimpse of the starry sky on a summer night and the four-dimensional continuum that has replaced our perceptual Euclidean space.
>
> In trying to distinguish appearance from reality and lay bare the fundamental structure of the universe, science has had to transcend the "rabble of the senses." But its highest edifices, Einstein has pointed out, have been "purchased at the price of emptiness of content."

Human science no longer experiences the world through human senses. Indeed, much of scientific knowledge these days completely contradicts our senses, which is why it is so difficult to accept such concepts as quantum mechanics and curved space. The sights and sounds and objects and motions around us are not divided up into small quantum bits, like still frames from a moving picture. The space around us does not seem to bend or change or curve. "This has led to a curious situation," writes Gregory. "The physicist in a sense cannot trust his own thought." And yet, he points out, we have to learn to live with the "nonperceptual concepts of physics. We are left with a question: how far are human brains capable of functioning with concepts detached from sensory experience?"

The answer has to be that there is more than one valid way of seeing things. If we listen to Bach (or Blondie) with our ears, and then "listen" to them again with (other kinds of) electronic detec-

*See "Waves and Splashes."

tors, we will pick up very different sets of signals. Both kinds of perception are equally "indirect." The difference lies mainly in the kinds of detectors we choose to pick the signals up. If we photographed the image that appears inside our eyes when we "see" something with an ordinary camera, we would indeed "see" something very different. There are many different windows on reality.* One of Einstein's most radical notions had to do with the multiple realities of space and time. That is, the space or time we perceive depends on the means we use to mark it off, on our point of view. "Space has no objective reality except as an order or arrangement of the objects we perceive in it," writes Barnett, "and time has no independent existence apart from the order of events by which we measure it."

Perception, that is, is a very *active* process. We do not just sit around waiting for information to rain down on us. We go out and get it. In the process, we alter it and even create it. One of the strangest things about the way physicists "see" elementary particles is that they often actually create them out of the energy of other particles to make them visible—something that doesn't seem quite "fair." But as Philip Morrison points out, you can't see the rapidly rotating blades of a fan unless you stop them or throw a rock at them. You can't sense radio waves by putting your hand in front of them, but you can if you tune your receiver so that it vibrates in resonance with the incoming signal. Even after people knew that visible light vibrated in waves, they couldn't see X-ray "waves" because the detectors were too large and clumsy. Only when someone had the bright idea of shining X-ray light through the thinly aligned layers of molecules in a crystal did the wave patterns emerge clearly. But the patterns—like the sound waves coming from the radio—are as much created in the process of detecting them as are the particles created in accelerators.

What we see is largely determined by the kinds of detectors we choose to turn on our world, by what we look for. This is true of everything from quarks and quasars to beauty and intelligence.†

It is also depends on our point of view. A house viewed from an airplane does not look at all the same as a house viewed from its

*See "Natural Complements."
† See "Measure for Measure."

own front door, or from the window of a rapidly passing car. (And the grass, as everyone knows, looks greener . . .) A baby does not recognize a toy viewed from the top as the same toy that looked so very different when seen from the side. A rotating shadow of any three-dimensional object will take on an amazing variety of different shapes. Which is the "true" perspective? It may be that the only wrong perspective is the one that insists on a single perspective. For physicists have found that light (like X rays, and even all energy and matter) can take the form of waves or particles, depending on how you look at them. Unfortunately, we often fail to recognize even quite familiar things when we see them in fresh

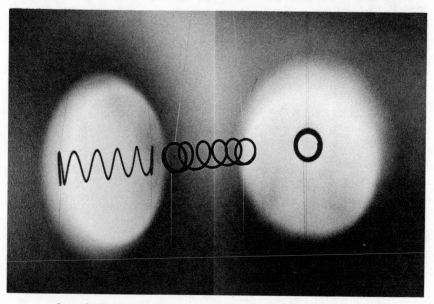

One shadow of the spring is a circle; another shadow of the same spring is a wave. If you could perceive the spring only by looking at its shadows, you might easily come to the conclusion that it was two different things.

perspectives, or unexpected contexts. Like the baby with the toy, we may be mistaking one thing for two. Space and time, energy and matter, waves and particles, are all different aspects of the same thing.

Part of the problem comes from our propensity to place boundaries between things where boundaries may not naturally exist.

Our eyes (like our minds) accentuate edges. There is a wonderful illusion in The Exploratorium that consists of a large, whitish board divided in half by a bunch of removable twine the staff calls The Horse's Tail. The board appears to be of one, even color. When you lift the "tail," however, lo and behold, the board separates into two squares of two different shades of gray—one quite a bit darker than the other. Yet the even more amazing truth is that the two sides of the board *are* identical, something your brain sees

The Horse's Tail (a.k.a. Gray Step 1): If you put your finger or a pencil over the edge dividing the two squares, the apparent difference between them disappears.

quite clearly as long as it's not distracted by the edge. Each square gradually increases in brightness from left to right. Because there is no sharp boundary *within* the squares, however, we cannot see the shading. Because there is a boundary *between* the squares, we exaggerate fiercely—and even extrapolate the difference at the edges to apply to the whole square.

Psychologists say that this is analogous to the way we often fail to see the large spectrum of abilities and behaviors *within* individual people, races, sexes, or nations—and yet we are quick to exag-

gerate the differences *between* them. The division into black and
white is something that takes place primarily in the mind's eye.
The forces and forms of the physical and biological universe are
ever-changing shades of gray. It is we who decide that the reptiles
are in a different class from the mammals, that space is separate
from time, that we cannot simultaneously think logically and feel
deeply. Of course, it is necessary to make distinctions in order to
see anything at all. The shape of things is determined largely by
their contours. But this kind of categorization easily leaps out of
bounds.

*A tiny meteor the size of a pea falling through the atmosphere
looks the same size to us as a giant star millions of times more
distant; for that reason, we call it a shooting star.*

Gathering, arranging, and sorting the information from the
outside world is only the first step, however. We still must decide,
What is it? What does the information mean? Meson or proton?
Streetlight or moon? Shadow or burglar? Planet or star? A small
dim object close up looks the same as a large bright object from

afar. Stick out your thumb and use it to size up a house far away outside the window. The house may seem no bigger than the tip of your nail. Then how do you know it isn't? "When a tiny meteor smaller than a pea is falling through the air," writes Jeans, "it will send the same electric currents to our brains as will a giant star millions of times larger than the sun and millions of times more distant. Primitive man jumped to the conclusion that the tiny meteor was really a star, and we still describe it as a shooting star."

You have no way of knowing that there is anything behind you until you turn around, and yet we are not surprised to see things there. You have no way of knowing that the next step will fall on solid ground, and yet you take it on faith. Sometimes we are

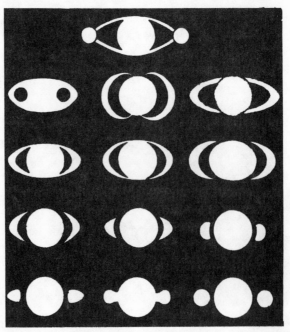

Richard Gregory offers these drawings of Saturn by Huygens as evidence that the early scientist did not recognize the curious patterns as rings—primarily because he did not expect to see rings.

fooled: the piece of fluff floating a few feet away looks like an airplane miles up in the sky (or vice versa). Yet most of the time we are remarkably accurate in the way we size up our familiar world and get around in it.

"Familiar," however, is the key. As Gregory points out, perception is a matter of seeing the present with images stored from the past; it is a matter of selecting the most likely (that is, the most familiar) object, the most "obvious" answer for the question What is it? "This acceptance by the brain of the most probable answer implies a danger: it must be difficult, perhaps somewhat impossible, to see very unusual objects." And if perception is a matter of making sense of the world with our limited collection of answers from the past, he asks, "then what happens when we are confronted with something unique?"

The answer is, we don't see it. The seventeenth-century Dutch scientist Christiaan Huygens drew detailed pictures of the planet Saturn as seen through his homemade telescopes. But he never recognized the unusual patterns as the now familiar rings. He couldn't see the rings largely because he wasn't *expecting* to see rings. "We not only believe what we see," writes Gregory, "to some extent we see what we believe. . . . The implications about our beliefs are frightening."

Gregory is hardly the only scientist to come to this conclusion. Loren Eiseley writes: "Each man deciphers from the ancient alphabets of nature only those secrets that his own deeps possess the power to endow with meaning." Sir Peter Medawar, in *Advice to a Young Scientist*, dismisses the notion that scientific discoveries are made by "just looking around":

I myself believe it to be a fallacy that any discoveries are made in this way. I think that Pasteur and Fontenelle would have agreed that the mind must already be on the right wavelength, another way of saying that all such discoveries begin as covert hypotheses—that is, as imaginative preconceptions or expectations about the nature of the world and never merely by passive assimilation of the evidence of the senses. . . . The truth is *not* [emphasis his] in nature, waiting to declare itself . . . every discovery, every enlargement of the understanding begins as an imaginative preconception of what the truth might be.

Sometimes it takes a long time before our innate preconceptions make way for a clear view of things we don't expect to see. Aristarchus saw in the third century B.C. that the earth revolved around the sun, yet this information wasn't assimilated until two millennia later, when Copernicus "discovered" it for the rest of us. Even Copernicus "saw" all celestial motions as perfect circles, when in fact the planets orbit in ellipses. It takes time to get used to social "discoveries" too—to build cultures and civilizations. Today it is generally known that black people as well as white people can be physicists. Yet very few members of minority groups are physicists, probably in part because the image we "see" of a physicist is a white man—much as we "see" the sun rising and

We see what we expect to see.

setting. For a girl or a black child to imagine herself as a physicist requires the same leap of faith that it requires to see the sun's "motion" as the result of the earth's rapid spin.

In fact, it is often hard to disentangle the image of the "real" world from the preconceptions we project on it. How much about the universe do we actually discover, and how much do we impose? Often, it is not clear which we are doing—like the person who heard a lecture on astronomy and afterward complained that while the lecturer had beautifully explained how the scientists had discovered the sizes and temperatures of the stars, he had neglected to tell how the scientists discovered their *names*.

Even experienced editors misread this sign because they see what they expect to see.

Douglas R. Hofstadter, author of *Gödel, Escher, Bach,* wrote in a recent issue of *Scientific American* about the dangers of perceiving things through the veil of expectation that colors our world so vividly in reflections of familiar images. In a piece entitled "Default Assumptions," he considered at length an issue that many people consider quite trivial—the established custom of using the pronoun "he" to stand for "he or she." Yet Hofstadter shows how even this seemingly minor semantic habit has a strange power to skew our senses.

He starts with a joke. It is the old riddle about the father who was killed in an auto accident, and the son in the seat beside him who was badly injured. The son was rushed to the hospital, but the surgeon who was to operate on him "on seeing the boy . . . blanched and muttered, 'I can't operate on this boy, he's my son.'"

The answer to the riddle, of course, is that the surgeon is a woman, the boy's mother. As Hofstadter tells it, he heard the riddle first "among a group of educated, intelligent people, some men, some women. . . . A couple of them, even after five minutes of scratching their heads, still didn't have the answer." He goes on to point out that our ability to make these assumptions, "to ignore what is highly unlikely—without even considering whether or not to ignore it—is part of our evolutionary heritage, coming out of the need to size up a situation quickly but accurately. . . . But the critical thing about default assumptions (well revealed by this story) is that they are made automatically, not as a result of consideration and elimination." The assumptions literally "blind" us to other ways of seeing things. Hofstadter goes on for another several thousand well-chosen words or so to document the point.

Perceptions are at best good guesses. They are the shortcuts we need to keep the information coming in from the outside world from becoming overwhelming—but they are never (necessarily) the truth. We have to judge a book by its cover because we don't have the time or the resources to look inside every book. We assume that if we see a face looking out of a window that there is a body attached to it, and that something that looks like a tree or a sailboat or a star probably is. We recognize patterns that seem familiar in a wide variety of contexts—even a child seems to know instinctively that a Saint Bernard is a "woof-woof" along with a Pekingese. But it is important to remember that it is we who have decided that these two very dissimilar creatures are both "dogs," in the same way as we call both the tool in the kitchen and a configuration of stars a "dipper." It is we who see the "man" in the moon. Snap judgments are useful and essential at the same time as they are treacherous and misleading.

"We cannot . . . believe in raw data of perception and suppose that perceptually given 'facts' are solid bricks for basing all knowledge," writes Gregory. "All perception is theory-laden."

We see what is familiar and we see what we choose to see, which are often the same thing. If people often engage in conversation only to disagree later completely on what was said, then in fact they probably really "heard" very different things—in the same way that different people can truly see the same event very differently. An astronomer who looks through a telescope at Saturn today has no trouble seeing rings. It depends on what you are looking for.*

A friend told me of a recent visit by her ten-year-old niece and her twelve-year-old nephew. "My niece immediately noticed all the dust balls in the corners of my living room," she said, "but my nephew didn't even notice that he'd spilled Coke all over the refrigerator." One woman defined liberation to me as that time when men would not only give children baths, but notice that they needed them. On the other hand, there are women who never notice that the car is out of gas.

How much we learn to see what we see struck me recently as I was watching a sonogram taken of the insides of my own body. The doctors and technicians hovered around, discussing what they "saw" as if it were as plain as the nose on my face. I couldn't see a recognizable thing. Some people can look at charts or equations or musical notations and immediately perceive a wealth of information that is completely invisible to the uninitiated. One of the first times I went sailing, someone kept telling me to watch for the dark spots on the water that told where the wind was blowing. Try as I might, I could not see a "dark spot" anywhere. Now that I sail more often, I find it much easier to pick out the dark spots and am often at a loss to understand why other people can't see them too.

(My friend the physicist told me of a time when, as a child, he was walking with his family in the mountains. Suddenly, everyone stopped and called, "Look at the deer!" But try as he might, he could not see it. His brother, mother, and father all pointed to the exact spot in the bush, told him what color to look for, and proceeded to describe the deer in great detail "Finally I saw it," he said. "And I realized that the problem was I had imagined a deer to be something much bigger than it actually was. Therefore I couldn't see it even though it was standing practically in front of my nose.")

What we learn to see is culturally conditioned—and this applies

---

*In *The Intelligent Eye*, Richard Gregory reproduces Christiaan Huygens's drawings of the planet Saturn as evidence that he did not "see" the rings. In *Cosmos*, Carl Sagan reproduces other drawings of Saturn, also by Huygens, which seem to show that he did indeed "see" the rings.

even to the things we see through the supposedly "objective" eyes
of science. When people first look through a microscope, they often
have trouble seeing anything but their own eyelashes, or random
specks of light. You have to be taught how to see an amoeba—just
as you "have to be taught" racial prejudice, as the song in *South
Pacific* says it.

In fact, Stephen Jay Gould writes of the continuing scientific
controversy about the superiority of certain races, and concludes
that there has never been anything "scientific" about it:

> Clearly, science did not influence racial attitudes in this
> case. Quite the reverse: an a priori belief in black inferiority
> determined the biased selection of "evidence." From a rich
> body of data that could support almost any racial assertion,
> scientists selected facts that would yield their favored con-
> clusions according to theories currently in vogue. There is, I
> believe, a general message in this sad tale. . . . Scientists
> tend to behave in a conservative way by providing "objec-
> tivity" for what society at large wants to hear.

In several other contexts, he shows that much of what is supposed to be science is really social prejudice, the problem being (as
Hofstadter pointed out) that "a bias must be recognized before it
can be challenged. Common sense is a very poor guide to scientific
insight for it represents cultural prejudice more often than it reflects the native honesty of a small boy before the naked emperor.
. . . The great geologist Charles Lyell argued that a scientific hypothesis is elegant and exciting insofar as it contradicts common
sense."

Science not only shapes culture,* it also is deeply, often invisibly, imbedded in culture; it needs to break out of that often un-
yielding ground before it can see the light. Culture is our spider
web, and a sticky one at that. Its influence is so strong that when a
brilliant supernova appeared during the height of religious belief in
Europe in 1054, not a single chronicle in Europe mentioned it—
even though it lasted for four months. It was not recorded because
it was not considered important. It was filtered out of history as

---

*See "Sentimental Fruits."

effectively as the bonging of my clock was filtered out of my living room.

Galileo first stepped into trouble when he had the temerity to see a very bright star that we now call a nova. "The appearance of a new star in heaven," writes Gamow, "which was supposed to be absolutely unchangeable according to Aristotle's philosophy and the teachings of the Church, made Galileo many enemies among his scientific colleagues and among the high clergy." Galileo's telescope revealed (among other things) that Venus and Mercury, like the moon, sometimes have crescent shapes—implying that they

The Exploratorium's Distorted Room is convincing evidence that people would sooner perceive children growing and shrinking to impossible sizes than a room without right angles.

orbit around the sun. But what he saw was "certainly more than the Holy Inquisition could permit; he was arrested and subjected to a long period of solitary confinement."

We need not go back in history to see examples of this cultural perceptual conditioning. Who would think that civilized people are so culturally accustomed to rooms with right angles that they would rather see a person shrink before their eyes than a room change shape? Yet this happens in the famous Distorted Room invented by the painter Adelbert Ames. The room is misshapen in such a way that a person viewing it through a thin slot (which makes binocular depth perception impossible) has to make a perceptual choice: either people inside shrink and grow to impossible sizes as they move about, or else the room isn't the usual rectangular shape. People always choose the right angles. Even our perceptual belief that things in the distance don't actually shrink is to some extent culturally conditioned. A psychologist I knew used to tell the story of a forest-dwelling tribe of pygmies who had no experience seeing things at long distances. One day, this tribe happened upon a herd of cattle, or some such animal, on a far-off

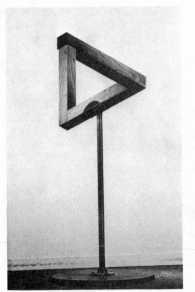

The Impossible Triangle appears impossible only because we insist on seeing it in the familiar shape we call a triangle. What is truly impossible is the ability to perceive totally unfamiliar objects.

plain—and of course assumed that they were tiny creatures, perhaps the size of ants. Imagine the pygmies' terror when the tiny creatures grew to great size as they approached!

It's easy to laugh at the pygmies, and it's fun to be amused by optical illusions. But illusions can be disillusioning. Usually the only illusion involved is a departure from our usual point of view, a contradiction of our preconceived notion of reality. Perceptual expectations not only make it impossible to see the shape of things correctly, they can lead us to see "impossible" things. A case in point is the Impossible Triangle, a structure of three thick beams joined at right angles. Of course, no triangle can consist of three right angles, but the Impossible Triangle is no ordinary two-dimensional triangle. It is a configuration of three beams joined at right angles that *looks* like a triangle from a particular point of view. It is only "impossible" as long as we insist on seeing it as a familiar triangle, instead of the unfamiliar figure that it really is. The unfamiliar figure, on the other hand, appears impossible, *"although it exists,"* says Gregory (with emphasis!).

We dismiss the real as impossible, and experience the impossible even though it cannot be real. If you soak one hand in ice water and one hand in hot water, then plunge both hands into a bowl of warm water, one hand will feel cold and the other hot—even though they are both resting in water of the same temperature. On the other hand, if you touch metal and Styrofoam at the same temperature, the metal will feel much colder. What you perceive is "impossible" only because your understanding of temperature is incomplete.

Intellectual knowledge alone does not change what we perceive. The Impossible Triangle, the Distorted Room, the experiment with the hot and cold water, retain their power to fool us even after we have learned how they "work." We still see the sun "rise" and "set." We see the stars as if they hang from a flat ceiling, and we can see that the earth is flat. We see the moon as a disk about a foot across and a mile away, even when we know it is a sphere about 240,000 miles away and 2,000 miles in diameter. Look at the sun sometime and try to see it as a star 93 million miles away. It is truly impossible.

The information we receive from the outside world may be ambiguous, but our perceptual prejudices are not. It's all too easy to believe in the rising sun and right angles in the same way as many people believe that a bleached blonde is a dizzy broad, that the man in the Brooks Brothers suit has a button-down mind, that women are not surgeons. They may even "believe" these things when they "know" better.

"It is an effort to get perceptual and intellectual knowledge to coincide," writes Gregory. "If the eighteenth century empiricists had known this, philosophy might have taken a very different course. No doubt there are also implications for political theory and judgment."

The moon in particular is a "heaven-sent object for perceptual study," says Gregory. The size and distance error in our perception of it are about a millionfold. But what is truly surprising, he says, is that we attribute any size or distance to it at all. We have

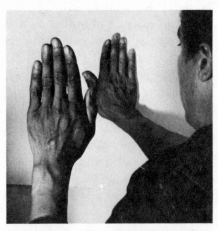

A camera is much more objective than the eye: it sees that a hand twice as far away makes an image half as big. Human brains, on the other hand, automatically compensate by making the distant hand appear "bigger." This means that the people and objects around us don't seem to shrink as they fade in the distance; but it also might explain why the moon looks larger on the horizon.

no frame of reference. We have no experience with objects so large or so far away. Then how do we know the brightness of stars, the size of quasars, the distance of far-off galaxies? The answer is, we make long chains of assumptions. If even one of those assumptions is just a little bit wrong, our perceptual conclusion will be a great deal wrong.

In fact, the moon, as almost everyone knows, changes size in the sky as our assumptions about its distance away from us change. The moon looks larger on the horizon than it does when it's high in the sky. There is no widely accepted explanation for this, but one theory is that the "ceiling" of the sky seems closer to us than the far-off horizon. Therefore the same size object, if it were closer, would have to be smaller to make the same size image in our eyes. A basketball across the room and a Ping-Pong ball at arm's length make the same size image in your eye. If you didn't know the size of the basketball, you would automatically "see" it as larger than the Ping-Pong ball merely because it was farther away. When we assume the moon is farther away on the horizon, we "see" it as larger, too.

In addition, we all have a wealth of experience that tells us that things that fly by us—planets, balls, planes, birds—all get smaller as they recede into the distance. If they *didn't* seem to get smaller, then it could only be because they were actually growing bigger! The same thing happens with the moon. It doesn't really grow bigger. But it does seem to recede in the distance. And since its image doesn't get smaller, you can only assume that it too is growing larger.

(If this seems hard to believe, you can prove it to yourself with a simple experiment. The next time a very bright light flashes in your eyes look at your hand, and you may see a small "afterimage" of the object floating on your palm. If you then look at a wall a few feet away, and then at a wall even farther away, the afterimage will get bigger and bigger. [It helps to blink to keep the afterimage in view; your brain tries to erase the "extraneous" image by pushing it aside, but blinking can bring it back.] The size of the actual image is imprinted on your retina, and always stays the same. But your brain automatically makes it appear larger or smaller, depending on the presumed distance away of the "object" making the image.)

One thing is clear, however. When you see the "large" room on the horizon, if you look at it upside down—say through your legs—so that the perspective of the horizon disappears, it will suddenly appear "small" again. So often we have to turn things upside down to see them in a proper perspective. A vacation to a strange place or trip to a foreign land can leave us with a totally fresh point of view. So can a book, or a play, or a film. And how quickly our perceptions change when we step into each other's shoes, take over each other's roles, see out of each other's eyes—if only for an instant.

There are still other limits to our perceptual powers: some of them have to do with the fact that we are a part of the nature we study (how clearly can you think about thinking, for example?); others have to do with the fact that if you look at something closely enough, you always have to disturb it. Behavioral scientists are constantly plagued by this problem, but so are nuclear physicists. There is no way to "see" the exact position and velocity of a sub-atomic particle because in measuring one aspect, you automatically disturb the other. In trying to look at subatomic things, we are, as Barnett says, somewhat in the position of a blind person trying to discern the shape and texture of a snowflake. As soon as it touches our fingers or tongue, it dissolves.

Some people have interpreted these perceptual limitations to mean that objective reality does not really "exist"—whatever that means. But the world is full of many important and lovely things that elude measuring. Love, for one. But also hate, humor, and almost all other emotions, ranging from our reaction to the first fresh smell of spring to our uncanny response to the arrangement of sounds we call music. If you try to dissect the emotional power of a painting bit by bit, it will dissolve (at least temporarily) as surely as a snowflake.

In any event, "reality" means different things to different people. As Nobel Prize-winning physicist Max Born put it, "For most people the real things are those things which are important for them. The reality of an artist or a poet is not comparable with that of a saint or prophet, nor with that of a businessman or administrator, nor with that of the natural philosopher or scientist."

This does not mean, Born says, that our sense impressions are some kind of "permanent hallucination." On the contrary, we can often agree on the nature of objective reality despite its many appearances: "This chair here looks different with each movement of my head, each twinkle of my eye, yet I perceive it as the same chair. Science is nothing else than the endeavour to construct these invariants where they are not obvious."

Our scientific perception of reality grows and gains confidence in the same way that a baby gains confidence in the everyday reality of his or her world. We not only make hypotheses, we *test* them (or we should). If the same things happen enough times, we gain confidence in our theories. If the rattle falls to the floor every time it rolls off the table, this is unlikely to be a coincidence. If what we see with our eyes is confirmed by our sense of smell or hearing or touch, so much the better. And if other people seem to sense the same things we do, the more convincing still. We cannot eliminate the subjective aspects of perception, but we can subdue them. "It is impossible to explain to anybody what I mean by saying 'This thing is red,'" writes Born, "or 'This thing is hot.' The most I can do is to find out whether other persons call the same things red or hot. Science aims at a closer relation between word and fact."

(Of course, we cannot always confirm what we see with other senses. We will never be able to travel to the stars to measure their temperatures, or analyze the elements that make them up. But as in everyday life, we can usually assume that what is most likely is true.)

Physicist David Bohm concluded that what we perceive is what is *invariant*—what does not change under altered conditions. The baby learns to perceive the bottle correctly when he or she learns that it does not change when it's viewed from various points of view. And it was the realization that the speed of light is invariant that led Einstein to his special theory of relativity. The surprising conclusion that space and time were relative came from the much more fundamental insight that the laws of nature (like the speed of light) were invariant under all conditions.*

The strength of our beliefs is buoyed by the strength of connections between our observations and beliefs. The more threads we can tie together, and the more tightly they are knit, the less likely it is that something major can slip unnoticed through a perceptual blind spot. Not only does the rattle fall to the floor, but the moon itself is falling toward the earth, and the earth toward the sun; rivers and rain and cold air all sink to the center of the planet. Gravity gains force as more and more things can be explained by it. Therefore even though there are many things that astrophysicists or subatomic physicists cannot really "see," even in theory, they gain confidence in their solutions to puzzles as more and more pieces fit. If some underlying idea were fundamentally wrong,

*See "Relatively Speaking."

writes Victor Weisskopf, "our interpretation of the wide field of atomic phenomena would be nothing but a web of errors, and its amazing success would be based upon accidental coincidence."

Our innate perceptual limitations are as necessary and frequently as reliable as they are sometimes deceiving, so there's no point in even thinking about getting rid of them. We will simply have to accept, as British astronomer Sir Arthur Eddington observed, that any true law of nature is likely to seem irrational to rational human beings. Still we can do what scientists do to diminish the margin of error: we can try to get to know our instrument, its calibration, position, limitations, frame of reference—in short, how it works.

That is, we can better get to know our perceptual selves. No longer is perception the relatively simple matter it was in the Galilean world where the universe was considered an object under inspection. We have reached the limit of that view, as Nobel Prize-winning chemist Ilya Prigogine points out in *From Being to Becoming:*

> To progress further, we must have a better understanding of our position, the point of view from which we start our description of the physical universe. This does not mean that we must revert to a subjectivistic view of science, but in a sense we must relate knowing to characteristic features of life. . . . In the words of Niels Bohr, science is a fascinating adventure in which we are "both spectators and actors."

The greatest danger is that the necessity to focus on the familiar will make it all too easy to miss novelty. As one scientist put it, sifting makes information, like processed flour, easier to use, but it also takes out the nutrients. Because we tend to screen out all that is different or threatening, we also screen out much that is illuminating. (The danger of this kind of screening may be even greater when computers do the sifting for us.)

Our optical illusions prove to be fairly innocuous compared to our personal and political ones.

3. FORCES, MOTIVES, AND INERTIA

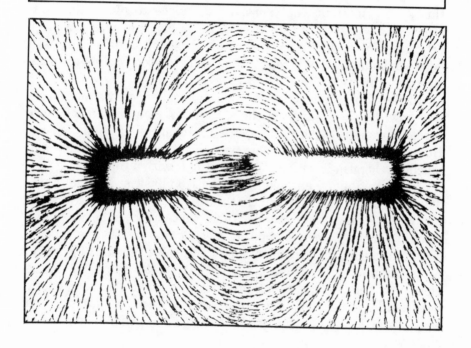

Nowhere have the words of physics infused the everyday vernacular so much as in the realm of forces—the influences that propel people to do things, or that leave them perpetually stuck in ruts. We speak of people who gravitate toward certain interests, who are pressured to succeed or bogged down by inertia, who cause friction in groups or organizations. We describe people as magnetic, forceful, repulsive, or even electrifying. We talk of being pushed into doing things, of being attracted by places, people, or jobs. We even talk of "pushy" people. Most of the time we think we know what we are talking about.

Physicists, on the other hand, tend to be much more cautious when speaking in terms of forces. Or as Caltech physicist Richard Feynman writes in his famous *Lectures on Physics:* "If you insist upon a precise definition of force, you will never get it!" (Exclamation his!) J. Robert Oppenheimer noted that even the great Newton—who probably came up with more formulas for forces than anyone else—never understood what a force really was. "Was this . . . something that spread from place to place, that was affected only instant by instant, point by point; or was it a property given as a whole, an interaction somehow ordained to exist between bodies remote from each other? Newton was never to answer this question."

In Newton's time, the great philosophical mystery about forces was how did they get from there to here? How did an influence spread through distances? Especially distances in empty space? How could you make something move if you couldn't reach out and touch it? Or push it with a stick? Or pull it with a rope?

In the mid-nineteenth century, British physicist Michael Faraday came up with the notion that these forces might be transmitted by something akin to rubber tubes that stretched between two poles of a magnet, or two opposite electric charges; he even speculated that a similar mechanism might transport the pull of gravity. Scotsman James Clerk Maxwell formulated these rather vague ideas into the precise equations that described electromagnetic force fields. A field is a kind of extended aura that surrounds a body, spreading its power to affect things like political spheres of influence.

Today, physicists are more likely to talk of forces as carried by special kinds of particles. These force particles often materialize out of the bursts of energy created when other particles collide in giant accelerators. The familiar photon, or light particle, is just such a force particle. But all this talk of "force particles" makes one wonder whether there is any difference anymore between the actual stuff of matter and the pulls and pushes that make it come and go.

In 1983, two of these force particles—the W and Z particles

that carry so-called weak force involved in radioactivity—were found at the European Laboratory for Nuclear Research (CERN) in Switzerland. The discoveries were hailed in the front pages of *The New York Times*, and were widely discussed in popular science magazines. But to people who are familiar with the forces of wind or tide, of springs and hammers, of fire and chemistry, of muscles and jet propulsion, of friction and magnets and gravity and static electric charges, this all seems very puzzling, to say the least. Or as someone finally said to me in total frustration after reading one of these articles: "I wish someone would explain these forces to me in terms of the force I feel when I stub my toe."

Unfortunately, while people certainly know a great deal about what forces *do*, they know a great deal less about the mechanisms that make them work. Explaining the what is easy; untangling the hows and whys is a tricky thing.

Take inertia, for example. In physics, inertia is simply a resistance to a change in motion—a kind of "hell no, we won't go" of matter. It's harder to toss a bowling ball than a tennis ball because the bowling ball has much more inertia. Inertia means an unwillingness to be pushed around, a proclivity to keep right on going the way you were going, or if you were stopped, to stay stopped. You can call it a habit, or a rut. But whatever you call it, it's not strictly a force. Newton defined force as an action exerted on a body in order to change its state; he said inertia was a measure of the *resistance* to that change in state. In this sense, force and inertia seem to be opposites.* Yet inertia feels exactly like a force when your car, for example, comes to a sudden stop and the "force of inertia" keeps you going through the windshield—or when you try to change the course of people or countries and come up against the tremendous inertia of history. Force and inertia may seem opposite in their origins, but they are often identical in their effects. A motive is anything that explains why we move. It does not matter whether the motive results from internal inertia or is exerted upon us by an outside force.

Newton referred to inertia as the "innate force of matter." The

---

*Of course, Newton also recognized that this resistance was itself a powerful force in his famous law of action and reaction. See p. 86.

more matter there is, therefore, the more inertia it naturally har-
bors. It's easier to push a light car than a heavy truck, and it's
easier to make changes in a small company than in the Defense
Department or IBM. You can pull the tablecloth right out from
under the table settings because inertia keeps the heavier glasses
and silver anchored in space on the table. This even has a parallel
in evolution: change tends to occur most rapidly in small, isolated
environments, says Stephen Jay Gould. "In large central populations, on the other hand, favorable variations spread very slowly,
and most change is steadfastly resisted by the well adapted population." And it also has to do with something called structural inertia, which depends on the reinforcing interactions of all a system's
parts. Large bureaucracies (somewhat like diamonds) are hard to
change not only because they are massive, but also because they
are complex and because they have long histories locked into them.

For all its obvious force, however, inertia itself is not that obvious. In fact, people often get easily discouraged when they try to
make major changes in social trends and public policy because they
underestimate this unexpected power of inertia. It took a rather
brilliant insight on the part of Galileo to recognize inertia at all.
Previously Aristotle (among others) had assumed that the natural
tendency of things was to come to a stop, and that planets (among
other things) needed a constant push to keep them going. The difference between Aristotle's view and Galileo's view was that Aristotle thought the natural state of things was at rest. Galileo, on the
other hand (perhaps anticipating today's workaholics), realized
that this natural state could also be motion.

Aristotle's was a very natural assumption. After all, cars, people, and even runaway roller skates eventually come to a stop if
someone or something doesn't supply the energy to keep them
going. In fact, *everything* in the everyday world eventually settles
into a "natural" state of rest. What Aristotle didn't realize was that
there was a hidden force behind that tendency to stop: the force of
friction. (And if you consider friction to be part of the natural state
of things, then Aristotle was right after all!)

If people and trends also seem so often to stop in their tracks,
however, it is not necessarily the natural order of things. It may
rather be because other things (or people) are rubbing them the
wrong way, sapping their energy, and slowing them down. The

only reason there aren't perpetual motion machines is that this contrariness exists almost everywhere in the universe, even in outer space, where only one atom fills each cubic centimeter.* If it weren't for friction, things would keep right on going until something came along to stop them, or turn them around. And it was this material stubbornness that Galileo recognized as inertia.

Newton later amplified the idea to include the notion that anything which *didn't* keep right on going required some kind of force to stop it. He realized that the moon would keep right on going— literally flying off on a tangent—if some force didn't pull it toward the earth. That force, of course, is gravity. But what is the force behind the tendency of things to keep right on going? The force that makes the moon (or anything else) want to fly off on a tangent, or a yo-yo to fly off in a straight line when you whirl it about your head and then suddenly let it go? The truth is, nobody knows. "The motion to keep the planets going in a straight line has no known reason," writes Feynman. "The reason why things coast forever has never been found out. The law of inertia has no known origin." Good old inertia turns out to be one of the deepest mysteries of nature.

There is a clue, however, to the source of inertia. And that clue comes from the famous (if almost certainly untrue) story of how Galileo climbed to the top of the leaning tower of Pisa and dropped two balls: a heavy one and a light one. The heavy ball and the light ball hit the ground at practically the same time. Even if the story is false, its lesson is true. Many science museums have exhibits consisting of vacuum tubes that allow visitors to watch firsthand as a feather and a coin fall to the bottom at the same rate. This doesn't happen in air, of course, because the air pushes back more on the relatively large surface of the feather than on the relatively small surface of the compact coin. But you can amaze your friends and neighbors (or at least I amazed my six-year-old) by dropping a piece of paper and a small rock to see which falls fastest. (The rock, obviously, but *not* because it is heavier, as you will see.) Now do the experiment again, but this time crumple the paper into a tight

---

*The pulsing of electric and magnetic fields that make up light come pretty close to perpetual motion, however. By the time it prints an image on your retina, light from a star may have been traveling for several million light-years.

ball. Now the paper and rock should hit the ground simultaneously—proving that *weight* doesn't matter. It may be true that
"the bigger they come, the harder they fall," but no one ever said
anything about bigger things falling *faster*. This experiment probably wouldn't work from the tower of Pisa because the thick pillow
of air between the top of the tower and the ground would cushion
the fall, but it works fairly well in your average living room, and
perfectly well in a vacuum.

Why does it work? Heavy things and light things fall at the
same rate in a vacuum for the simple reason that while gravity
pulls harder on heavy things, heavy things also have more inertia—so they *resist* harder. Philosophically, this is somewhat de-
pressing. It means that you can never get anywhere (using
gravity) because the harder gravity pulls on something, the harder
it is to pull.*

Of course, gravity clearly pulls harder on big things: big people
weigh more than little people. But big people are harder to push
around, so it's impossible to get them going (falling) any faster.
This explains why a pendulum with a heavy weight at the end and
a pendulum with a light weight at the end will swing at the same
rate, and why a heavy object (like a satellite) will circle the earth
at the same rate as a light object (like an astronaut) inside it—
making the astronaut weightless. For both the astronaut and the
spaceship, the gravitational pull toward the earth is exactly balanced by the inertial tendency to fly outward. So if you drop your
pencil while whirling around the earth in a space station, you don't
have to worry about its falling to the floor. It will keep on orbiting
or "floating" exactly where it is.

Einstein thought it was a funny coincidence that gravity and
inertia should balance each other so perfectly. In fact, he didn't
swallow the coincidence at all. He looked for the reason behind the
coincidence and came up with the conclusion that gravity is not a
force but rather the geometry of the space we live in. This is the
basis of his general relativity, the idea of curved space. And the
"springboard" for this brilliant insight, writes Lincoln Barnett,
was nothing other than "Newton's law of Inertia which . . . states

---

*Although we shouldn't confuse cause and effect; see "Cause and Effect."

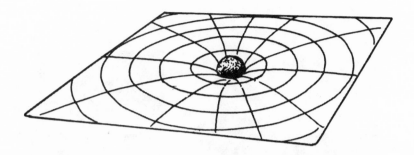

Einstein's General Relativity describes the "force" of gravity as the unseen geometry of space. In this two-dimensional analogy, space-time near a massive star is curved in a way similar to the surface of a water bed when a heavy ball rests on it.

that 'every body continues in its state of rest, or of uniform motion in a straight line, unless it is compelled to change that state by forces impressed thereon.'" Even the equation that ignited the atom bomb ($E = mc^2$)—and in fact ignites every match and star and candle—first appeared in a paper entitled "Does the Inertia of a Body Depend on Its Energy Content?" So inertia is not only mysterious; it is also rich and profound.

In fact, inertia may turn out to be nothing less grand than the combined gravitational pulls of all the matter in the universe. This idea was first proposed by the Austrian mathematician and philosopher Ernst Mach more than a hundred years ago, and no one (that I know of) has come up with a better explanation yet.* It imagines inertia to be the result of everything pulling on everything else—a well-known phenomenon that tends to get people stuck, say, in the center of a milling crowd. Indeed, I suspect that most of the time when people get stuck in certain places in their lives, it's not so much because they lack the internal energy or outside forces to move them along. It's rather because forces are pulling them in so many conflicting directions. If a single force comes along that is strong enough to overwhelm all the others,

*See "Relatively Speaking."

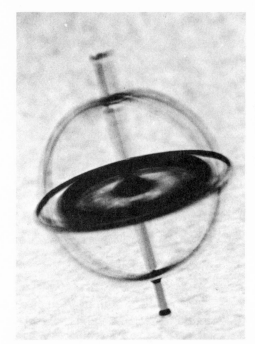

Anything that spins has a built-in resistance to tipping—be it a top, an electron, or a bike wheel.

then it suddenly becomes very simple to overcome inertia and escape most common ruts.

So questions of the origins of inertia really boil down to questions about how matter in one part of the universe communicates with matter in another part of the universe. Because if a particular rock, say, has a certain amount of inertia, it means that it resists changing its motion with respect to the rest of the universe (you, the earth, and so on). So there must be some way for all the bits of matter in the universe to make their influence felt on each other. Whether they do it through gravitons or gravity waves or the warp of space itself, it certainly makes inertia a more intriguing thing than a simple inability to pull yourself out of bed in the morning.

Fighting against inertia really amounts to pulling or pushing off the stars. No wonder it sometimes seems so hard to change things!*

---

*The idea also has many problems, not the least of which is that if the gravity of

The first time I ever saw the phrase "pushing off the stars" was on an exhibit in The Exploratorium about gyroscopes. This connection will be immediately familiar to those who fly planes, because they often get around with the help of combinations of gyros appropriately known as *inertial* guidance systems. They work because of the curious fact that spin itself has a kind of inertia. If you spin a bicycle wheel and then try to tip it, you can easily feel the force of resistance. A spinning top stays upright because the inertia of the spin resists the pull of gravity. A top that's not spinning falls right over.

A gyro is simply a spinning wheel that resists a twist or tip to its motion in the same way as a large rock resists getting pushed around. If you set a spinning gyro in gimbals so that the pull of gravity is balanced and neutralized, it will continue to point in any position you put it in (say, north) regardless of the changing orientation of its environment (say, a moving plane). Of course, you have to be careful. A gyro keeps its orientation with respect to all the matter in the universe—not just the earth. It can't point to a fixed position on earth because the earth is always moving. So if you take off from New York and set your gyro compass for Chicago, Chicago won't necessarily still be where the compass is pointing by the time you get there.

But the important thing is that the *source* of the resistance to a change in the orientation of a spinning wheel is every bit as mysterious as the source of the inertia in the Rock of Gilbraltar. That's why The Exploratorium exhibit talked about "pushing off the stars." Because what else does a spinning wheel push off? How does it communicate its spin to the rest of the universe around it? In fact, in a sense, we are always pushing off the stars. Even to walk, you have to push off the earth. The earth is attached (gravitationally) to the rest of the solar system, just as the solar system whirls around with the rest of the galaxy and the galaxy we live in is a part of the total motion of the universe. If Mach is right, anything that resists the push of your step when you walk (say, the inertia of the earth) is really pulling on the stars.

---

far-off stars combines with the gravitational pulls of other things around you to produce inertia, then the gravity from the farthest stars would have to start traveling toward you before the gravity from nearby planets, and so on.

I still haven't answered my friend's question about "the force I feel when I stub my toe." In one sense, the force he feels is the combined gravitational pulls of all the matter in the universe—inertia. Except that inertia isn't considered a force. And the doorsill does more than merely *resist* a change in motion when you stub your toe. It actually pushes back. This is proven by the fact that stubbed toes are also squashed toes and painful toes. Something has exerted a force on them. In fact, the only reason you can walk at all is that the earth pushes forward on you and when you push back on it, and the only reason you can drive a car is that the road pushes forward on the wheels as much as the wheels push back on the road. The earth doesn't move backward (very much) when you walk only because it's much more massive than you are.

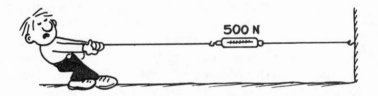

If you pull on a rope with a force of 500 Newtons, the wall pulls back on you with an equal and opposite force that you can measure on a scale.

This is known in physics as Newton's famous law of action and reaction: "Whenever one body exerts a force on a second body, the second body exerts an equal and opposite force on the first." Forces always come in pairs. For every action, there is an equal and opposite reaction. It doesn't matter which is which, and often it's hard to tell which is which. This is especially true when the law applies to people, where it is more popularly known as tit for tat, or "an eye for an eye," or "I'll get you for that." The difference is that in human terms, the reactions are rarely equal, but tend to escalate rapidly—every reaction having more retaliatory force than the initial action. What to one person is an action (say, supplying arms to a hostile country) is to another person a reaction (say, the attempt to prevent a revolutionary coup being provoked by the first person's country). And so on.

When it comes to countries, the law of action and reaction usu-

You cannot touch without being touched—Newton's Third Law.

ally gets you nowhere—as in the nuclear arms race, and policies of Mutually Assured Destruction (appropriately known as MAD). But when it comes to physics, it can get rockets to the moon and even outside the solar system. If you fill up a balloon with air and let it go, it races around the room because the air that is being pushed out by the walls of the balloon is pushing back on the balloon, sending it flying. In the same way, a bullet pushed forward by a rifle kicks back with a power that can knock you to the ground.

This "reaction force" is more obvious in the absence of friction. When you use the force of your muscles to bang on a door, it is not obvious that the door bangs back on you, because friction keeps both you and the door more or less in place. If you banged on a door while you were wearing roller skates, however, your banging could easily start you moving backwards. And if two people on roller skates toss a ball back and forth, they'll drift farther and farther apart, because each time they throw the ball, the ball will throw them a little bit backwards. The equal-and-opposite equation applies to all forces human and cosmic and mechanical. The earth pulls on an apple with a force equal and opposite to the pull of the apple on the earth, just as the equal and opposite pulls and pushes of light particles called photons account for electrical attractions and repulsions.

This cosmic game of tit for tat has the curious consequence that nothing can hit you back any harder than you hit it. Or as physics teacher Paul Hewitt likes to demonstrate in his classes, there is no way you can strike a piece of paper with an appreciable force. Hit it as hard as you like, and all you will ever feel is a slight tap. Since the sheet of paper cannot "hit back" with, say, fifty pounds of force, it means that you cannot hit *it* with fifty pounds of force. "You can't be hit harder than you can hit back," says Hewitt. "You can get only what you give. You cannot touch without being touched."*

I am reminded of the time an Exploratorium staff member proposed that the museum retaliate in some small way for an injustice or rudeness inflicted by another institution. And the director responded by repeating his oft-quoted injunction that "the worst thing a son of a bitch can do to you is to make you into a son of a bitch." Which is another way of saying that it's nearly impossible to do something nasty to somebody else without at the same time having it do something nasty to you, just as it is virtually impossible to confer a kindness without having it reflect nicely on you. There is a strange kind of justice in the philosophical application of Newton's Third Law. It implies that for every force you inflict on somebody else—for better or for worse—an equal reaction will return, like a boomerang, to inflict itself on you.

Of course, reaction implies a force—just as a resistance to change implies a force. But what kind of force is doing the reacting or resisting? We use the term "force" very loosely. Often, people say that they are forced to do something as a *reaction* to something

---

*Victor Weisskopf points out, however, that while it is frustrating to punch a marshmallow because the marshmallow doesn't punch back, you could with the proper equipment accelerate even a marshmallow to the speed of light, and you could hit even an electron with fifty pounds of force. Frank Oppenheimer cautions that even the "equal and opposite" isn't always true either—for example, in cases where you have stored energy, as in a mousetrap. If you kicked a piece of dynamite hard enough, you would certainly get back more than you put into it.

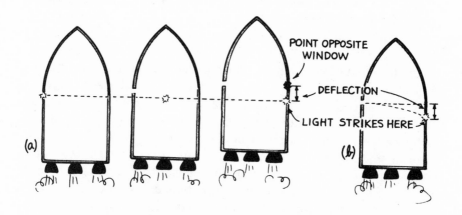

(a) An outside observer sees a horizontally thrown ball travel in a straight line, and since the ship is moving upward while the ball travels horizontally, the ball strikes the wall somewhat below a point opposite the window. (b) To an inside observer, the path of the ball bends as in a gravitational field.

done by somebody else. The force of habit is analogous to inertia. Physicists even recognize a whole category of forces as "fictitious" or "pseudo" forces. These pseudoforces are those that appear to be forces in one frame of reference, but not another, just as one person may say he was *forced* to behave in a certain way while to another person looking from a broader perspective it may appear that the first person is just following his normal behavior pattern. That is, whether or not you are actually *forced* to do something can be based on your point of view.

Say, for example, you are riding in a rocket ship that is accelerating for a fast trip to an outer planet. Suddenly, you drop your wallet. If your rocket were just drifting around in orbit, then your wallet would "float" in space, but since the rocket is accelerating, the floor soon overtakes the wallet and it appears to "fall." From the point of view of someone sitting on, say, a passing moon, it would be clear that the wallet remained relatively stationary while the accelerating rocket raced to catch up with it. But to those inside the rocket, it would appear that some outside force—gravity, or a magnet, perhaps—was attracting the wallet to the floor.

Either way, the wallet eventually hits the floor. "Whether a force is fictitious or real becomes purely a problem of language," as physicist James Trefil points out. "It is a totally irrelevant question as far as any physical effect is concerned." The falling wallet is similar to the kind of "thought experiments" Einstein used to develop the idea of relativity—which doesn't mean that "everything is relative" but that no matter how you view the situation, the physical outcome is the same. Gravity, according to Einstein, is just this kind of pseudoforce. But that doesn't stop the moon from orbiting the earth or apples from falling to the ground.

For that matter, magnetism turns out to be an *entirely* relative force. That is, magnetism is always created by a moving electric charge. The magnetism in iron magnets is caused by the spinning of countless electrons all twirling around in the same direction, just as the magnetic field of the earth is created (most likely) by the motion of electric currents deep inside its metallic core. Every time an electric current passes through a wire, it creates a magnetic field around it. Yet if you traveled along with an electron somehow, the magnetic force would seem to disappear, just as motion of an airplane traveling five hundred miles per hour seems to "disappear" when you're sitting in your seat watching an in-flight movie.* Magnetism is a relative effect of electricity, just as the force of "wind" on your face on a still day is a relative effect of riding a speedboat at sixty miles per hour. (Sailors, in fact, call this "apparent" wind—which is appropriately analogous to the physicists' "pseudo" force.)

FOUR FUNDAMENTAL FORCES

Reaction forces, relative forces, and even pseudoforces are always understandable in terms of at least one of the four so-called fundamental forces of nature. These forces are good old gravity (carried by as yet undiscovered particles called gravitons); electromagnetism (carried by particles of light or photons); the strong force (carried by gluons); and the weak force (carried by W and Z particles).

---

*Of course, the magnetic force doesn't just vanish; it reverts to electricity. Electricity and magnetism are two sides of a box. You can see it from one side, or the other, or a bit of both.

The choice of four is hardly arbitrary. It has resulted from a long series of sometimes startling discoveries about how forces work and how they relate to one another. For a long time, for example, people thought that the force of gravity was balanced by an opposite force called "levity," which caused things (like smoke) to rise. It took a great stroke of insight for Benjamin Franklin to realize that the static electricity that caused small-scale sparks around the house was also the stuff of lightning, and it wasn't until the nineteenth century that electricity and magnetism were recognized as different aspects of the same thing. No wonder the search for unity among forces has such an attraction for physicists. It has been so successful and revealing in the past that it is only natural to assume that it will continue to unravel fundamental mysteries in the future.

The most familiar force, of course, is gravity. It is the glue that keeps us stuck on the earth and pulls the elements of the earth in toward the center to form a compact sphere. It keeps not only our furniture from floating away, but also the air, the clouds, and even the moon. It makes the rain and baseballs fall. It is the force that everyone fights to push themselves out of bed in the morning, that their feet and legs stand up against all day. We grow up in response to the way gravity pulls us down, so gravity determines our shape—whether we are trees or children or elephants. Gravity is the major force behind tides, and weather, and floods. It is even at the bottom of black holes. One of the first great "unifications" of forces was Newton's insight that the force that makes things fall on earth is the same stuff that controls the shape of things in the heavens. Gravity is truly universal. It is also, as one physicist said, insatiable: while an electric charge can find a mate and become neutralized, gravity never lets go. And while there is a limit to the number of electric charges that an atom will attract, there is no limit to the amount of matter gravity can grasp toward a star.

In 450 B.C, Empedocles of Agrigentum speculated that the earth was made of meal, cemented together with water. He was very nearly right, and almost completely right if you take meal to mean matter and substitute "electricity" for "water." Electricity remains largely unnoticed until it flashes out at us during a lightning storm, but electrical forces are truly the "stuff" of matter.

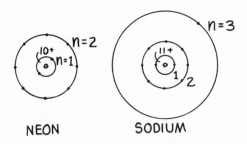

NEON   SODIUM

Neon's complete outer shell of electrons makes it chemically inert—a "noble gas" with little inclination to mingle with other elements. Sodium's extra electron, on the other hand, makes it eager to join with other atoms—especially atoms with empty spaces in their outer shells.

The nucleus of every atom is a tiny bit of positive electric charge, exactly balanced by the negative charges of its outer electrons. Since opposite charges attract, electricity is literally what holds atoms together. And since the configuration of outer electrons determines what attracts what, electricity accounts for all chemical reactions. An extra electron in the outer shell gives a substance the electrical conductivity it needs to carry a current; it also makes it eager to join with other kinds of atoms with "holes" in their electron shells. An atom with an extra electron in its outer shell, or with a "hole" or missing electron in its outer shell, is usually chemically "active." Atoms with complete outer shells have little interest in interacting with anything and are said to be "inert." Sodium (with one extra electron) naturally joins with chlorine (with one missing electron) to form sodium chloride, or salt. But neon, with a closed shell of ten electrons, is inert. And so on.

When you stub your toe, it is really the outer electrons in your toe atoms that collide with the outer electrons of the wood atoms; so electricity is "the force you feel" when you stub your toe. It is even electricity that makes it impossible to stub your toe in mush—because electricity accounts for all properties of matter: the hardness of wood, the transparency of glass, the glitter of gold. The interactions of those outer electrons buzzing around the atomic nuclei are responsible for everything from fire to thought; from cooking and digestion to taste and smell; from the solvency of water to the cleaning power of soap. Electricity is the force that makes sticky things stick together; it is behind the capillarity that pulls water up the trunks of tall trees and through the veins (and capillaries) of animals (including humans); it is even the source of friction. It is the basis of all terrestrial forces other than gravity.

Electricity becomes more impressive still when you consider that a moving electric charge produces still another force—magnetism. And that electricity and magnetism together form a continually alternating wave train that zips along through space at 186,000 miles per second, accounting for all radiation, including visible light, heat, microwaves, radio and television signals, X rays, and gamma rays.

The strong force and the weak force have remained hidden like genies inside the atomic bottle, springing into sight only recently. The strong force is sometimes called the nuclear force, for its realm is within the nucleus and its responsibility is nuclear reactions—most importantly, holding the constituents of the nucleus together. If there were no strong force, there would be no elements other than hydrogen with its single nuclear proton; there would be no planets, no life. The strong force is the force that fuels the nuclear reactor and the nuclear bomb, the sun and the stars. As physicists have seen further into the core of the nucleus, they have discovered that the nuclear force is probably a kind of complicated effect (like chemistry) of a still more fundamental force called the color force—which has nothing to do with visible color. It is the color force that is carried by gluons, and that acts between quarks. The nuclear force is to color what chemistry is to electricity.

As for the weak force, suffice it to say that it is the force behind radioactivity and certain reactions in the sun. Radioactivity is of no small consequence, of course. It has kept the earth warm enough to support life, and the random mutations it causes have helped along the evolution of all species. Of late, the weak force has been partly "unified" with electromagnetism in a display of mathematical wizardry reminiscent of Maxwell's unification of electricity and magnetism more than a century earlier. It was this recent electro-weak unification that pointed to the discovery of the W's and Z's—the weak force carriers—at CERN.

A fast look at these forces tells you immediately why physicists are so interested in unity. They seem to have nothing whatever to do with each other. Gravity, for example, works only one way. It

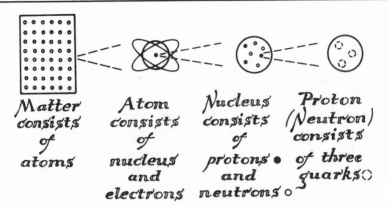

The different layers of the structure of matter.

ulls everything *toward* everything else, which is one reason that so many things in the universe are round. It also explains why gravity is so noticeable. In truth, gravity is *trillions and trillions* of times weaker than electricity. We don't normally notice this powerful electric force directly because in most of the universe it is balanced and therefore neutralized. That is, electricity, unlike gravity, both attracts and repels. All over the universe (and especially on the relatively calm and cool surface of the earth), bits of negative electric charge join with bits of positive electric charge to form bits of neutral matter. When you rub some of these charges off their neutral atoms—say, by shuffling across a carpeted room—and then allow them to get back together—say, by touching a doorknob—you can create small sparks. When the strong updrafts in huge thunderclouds rub raindrops together like your feet rubbing on the rug, huge numbers of electrons can be ripped off. When they get back together, they create the much larger sparks we know as lightning.

In other ways, gravity and electricity (or electromagnetism) are remarkably similar: both decrease in strength as they spread out through space according to the same equation, and both can reach—theoretically, at least—to the ends of the universe. The color force, on the other hand, seems to *increase* infinitely as it spreads out from the vicinity of a quark, or gluon. The farther two quarks drift away from each other, the more fiercely they are pulled back together. Thus, quarks are permanently trapped, and no "free" quark has ever been (or perhaps ever can be) found.

Odder still, the little-understood color force seems to disappear altogether at very close ranges—leaving the quarks free to rattle around within a tightly closed bag.

Forces not only act in very different ways, they also make themselves felt over very different ranges. While the gravity of the earth may be able to contribute to the inertia of a distant star (and vice versa), the extremely parochial weak force exerts its influence over a distance a thousandth the diameter of a proton. Gravity is both the strongest and the weakest force in the universe, depending on the range and scale you consider. It is a major force in the lives of universes, people, and stars, but virtually unfelt by atoms or even small spiders. The lives of lightweight things from small plants to molecules are ruled rather by chemical forces—surface tension, cohesion (stickiness), and capillarity—and all these forces are essentially electrical. The strong force is a major power within the nucleus, but dwindles almost to nothing once it leaves the confines of that nucleus. It has strong but very short-range hooks that don't come into play until two nuclear particles get very close together.

This last, at least, is explainable. That is, if you consider the nuclear force to be a kind of short-range effect of "color" in the same way that chemistry is a short-range effect of electricity, you can understand why the forces behave the way they do. Say, for example, you have a positively charged atom and a negatively charged atom, floating around in space. Their electrical influences reach, in theory, to the ends of the universe. But once they are attracted to each other and join, the resultant molecule becomes electrically neutral. Electrical forces are long-range, but chemical forces are short-range, because atoms usually have to get fairly close together before they can react chemically. The point of heating and otherwise cooking things to produce chemical reactions is to bring the atoms close together. Normally, their outer electrons repel each other, but heating gives them the extra energy they need to break those barriers and react.

On the subatomic level, the color force binds quarks together inside a proton or neutron in a way similar to the way the electromagnetic force binds the electron to the nucleus inside the atom. Inside the nuclear particle itself, however, opposite color changes cancel each other—leaving the proton or neutron "colorless." The color force is alive and pulling *inside* the proton or neutron, just as every atom is internally electrified. But these forces aren't normally felt on the outside, because they are canceled. Only when two atoms are very near can their internal forces reach out and touch each other, causing a chemical reaction. And only when two nuclear particles are very near each other can their short-range hooks grasp each other—causing, in this case, a nuclear reaction.

This interplay between short-range and long-range forces is not totally unfamiliar. Indeed, almost everyone can think of parallel ways it applies to human relations. Internal tensions within a marriage or a family (or even within a person, for that matter) are often completely unfelt by those on the outside, even though they may be explosive to those on the inside. And there is a very different quality between long-range attractions (a strange face across a crowded room) and the intimate bonds of long-term friendships. Many kinds of interpersonal forces—like physical forces—don't come into play until the participants get very close, whether they are the surprising explosive forces that can erupt in an angry crowd, or the unexpected attractions that sometimes take place when people are accidentally "thrown" together.

I can't help thinking, in this connection, of the well-known "long-range" repulsion that exists, for example, between sailboaters and powerboaters: "windbags" and "stinkpots." I was for a long time one of those sailing chauvinists who like only to ride on the wind and consider the people who whip around the harbor stirring up choppy wake and smelly diesel fumes a different breed—if not a different species. Then one day I was invited to ride on a supercharged seventy-five-mile-per-hour Hawk Excalibur racing machine, and was frankly awed by its beauty and the thrill of the ride. Today, I remain primarily a sailor. But because of my (since several) close encounters with speedboats, I can at least understand their powers of attraction. (The same is true of rock music fans and classical aficionados, of country dwellers and urbanites— even of believers of various religions. Of course, bringing mutually repulsive things together doesn't always result in closer bonds. Sometimes it causes explosions.)

Finally, different kinds of forces make their effects felt on different kinds of things. Particles that carry the electric force can

influence only electrified particles. Gluons carry color only between quarks and other gluons. The weak force is so specialized
that it interacts only with left-handed particles and right-handed
antiparticles (never mind what that means; the weak force is particular). But gravity affects *everything*. The source of gravity is
mass itself, and everything has mass if only in the form of energy.
(That is the meaning of Einstein's equation $E = mc^2$: energy equals
mass $\times$ c [the speed of light] squared.) Therefore gravity can even
bend a light beam skimming by a star. This universal property of
gravity is what allowed Einstein to see it as the grand geometry of
all (curved) space.

FIELDS AND PARTICLES

In a sense, curved space is simply the shape of the gravitational
force field that surrounds a piece of matter. A field is a kind of
tension in space that extends out around a particle (or a planet) like
the sticky spokes of a spider web, spreading its influence to all
other particles that might come into its path, and even sucking in
unwary particles like a vacuum cleaner. Field was the concept
used to overcome the main objection to Newton's ideas about gravity. Newton never answered the question How does gravity reach
out and grab the apple or the moon? How does the force get from
here to there? The concept of field eliminated the worry about "action at a distance" by temporarily eliminating the distance. The
field spread the influence of the particle out into space itself, much
as annexation can spread the influence of countries. (Exchanging
diplomats or fire would be analogous to exchanging "force-carrying
particles.")

At first, a force field was just an interesting way to look at how
forces behaved. That is, if you sprinkle iron filings near a magnet,
they will line up in a shape that corresponds to the magnet's force
field. In the same way, planets orbit the sun in a way that "lines
them up" with the sun's gravitational field. The notion of field geo-
metrizes forces, turning them into an integral part of the landscape
itself. Say, for example, you threw a ball down an invisible pipe
that was bent in several places. You could say that the ball was
"forced" to follow the bend by the pressure of the walls of the pipe,
or you could say that the ball naturally traveled in a bendy fashion

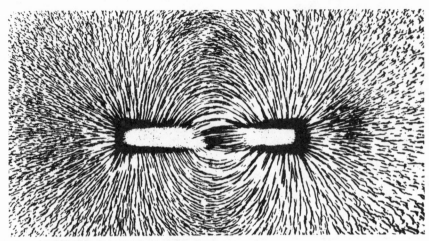

*Iron filings fall into shape along invisible lines of force sur-
rounding an ordinary bar magnet.*

because the space around it was curved. Large-scale force fields
(like gravity) are said to curve space, while small-scale forces (like
electromagnetism) are sometimes said to "wrinkle" it. (Actually,
they wrinkle the electromagnetic field.) Either way, a force field is
a useful mental and mathematical tool for visualizing how forces
operate. It is a description of the effect of a force which tells the
strength and direction of that force at every point in space.

It turns out, however, that force fields are much more than
that. Force fields exist *on their own*, independent of the particles
responsible for creating them. A planet or an electron is the source
of a force that creates a tension in the space around it. But that
force—that tension—can exist in the absence of the planet or electron. When a wiggling electron in the sun sets up a local wrinkle in
its surrounding electric field, that wrinkle speeds along at 186,000
miles per second and almost eight minutes later sets up a wrinkle
in an electric field somewhere on earth, where it might be detected
as "light." That wiggle takes time to get from there to here. If the
sun went out during the eight minutes it took the light particle to
travel to the earth, we would still be able to see it. When a supernova explodes, its gravitational effects can be detected years after
the event. Force fields have lives of their own.

The idea that a force field could be a separate entity was an
important first step toward the notion of forces as particles. The

second step was quantum mechanics—which grew out of the finding that everything, including the energy of force fields, is quantized, or comes in clumps. Therefore a force particle, like a photon,
is really a small clump of an electromagnetic field that travels from
place to place at the speed of light, carrying its quantized parcel of
energy and momentum with it. Force fields and force particles may
seem like very different things. But most of the differences lie in
our dearth of proper imagery.* Victor Weisskopf fought for years,
he says, to have force particles called "field quanta" to retain the
connection. Or as his colleague at M.I.T. Vera Kistiakowsky told
me: "It sounds confusing because you talk about it differently in
different situations. On a macroscopic scale, you use field; when
you get to the interaction of a single particle, you talk about
gluons. But it's one and the same thing." Field and particles are
complementary descriptions of the same phenomenon.†

Nevertheless, the imagery of force particles has become deeply
rooted in physics—and especially in popular writing about it.
You'll often read about particles being "exchanged" between other
particles like the ball that is "exchanged" between two people on
roller skates. The force of the exchange, says the imagery, is what
drives the particles apart. This imagery doesn't explain attraction,
however. In this case, two particles are drawn together by "sharing" the same force particle, much as two teenagers are drawn
together by sipping an ice-cream soda through two straws. In the
same way, two people can be drawn together by sharing experiences, or houses, or umbrellas. Exchanges of words can be used
either to attract or to repel. Many chemical bonds are the result of
a sharing of electrons among atoms.

If forces can be particles, however, is there any difference between matter and the pulls and pushes that make it come and go?
Is there any difference between "stuff" and "influences"? Can you
separate the actions from the actors? The things people (or particles) do from the people or particles themselves?

In physics, at least, you can. For one thing, matter particles
(like protons, neutrons, and electrons) obey what is known as the
Pauli exclusion principle, after Austrian physicist Wolfgang Pauli.

---

*See "Science as Metaphor."
†See "Natural Complements."

The notion of atomic electron shells rests on the Pauli principle because it says that no two electrons can occupy the same state. If all the spaces or states in shell number two, say, are full, then the next added electron must occupy a space in the next available

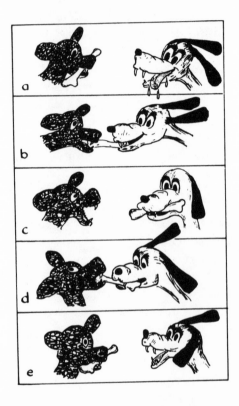

The exchange of a force particle between two nuclear particles brings them together.

shell. The Pauli principle explains why matter is not compressible, so it is really a "principle" that makes things solid and also the real force you feel when you stub your toe! This same principle (also known as electron pressure) is what keeps stars from collapsing. When a star is so massive that its gravity overcomes electron pressure, it literally collapses: the electrons compress into the nucleus, joining the protons to form neutrons and thus making a neutron star. If even the nuclear forces aren't strong enough to hold back gravity, the star keeps on collapsing—in theory at least forming what is known as a black hole.

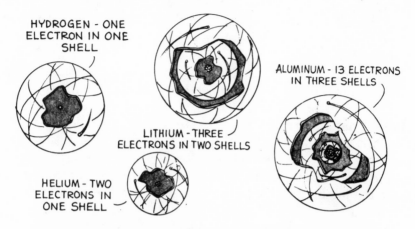

HYDROGEN - ONE ELECTRON IN ONE SHELL

ALUMINUM - 13 ELECTRONS IN THREE SHELLS

LITHIUM - THREE ELECTRONS IN TWO SHELLS

HELIUM - TWO ELECTRONS IN ONE SHELL

Atoms are mostly empty space, but they are nonetheless solid because only a limited number of electrons can orbit in each "shell." This electron pressure is the force you feel when you stub your toe.

Force particles, on the other hand, do not obey the Pauli exclusion principle. You wouldn't stub your toe if you tripped over a light beam. Force particles are described by a different set of statistics, called the Bose-Einstein statistics, which is why force particles are "bosons." Most matter particles are described by what are known as Fermi statistics, after Italian physicist Enrico Fermi, and so are called fermions. But there is also another difference between "force" and "matter," as Weisskopf likes to point out: fields are produced by particles and can be absorbed by particles, but the particles themselves are more permanent. If particles are objects and people, then fields are like words: they can be fleeting, but have long-lasting effects.

The many uncanny connections between matter and forces and fields may even lurk behind some of our confusions about how people affect each other. Is a forceful or attractive or repulsive person surrounded by a kind of "field" of influence? Or do they produce the effects they do because they are constantly sending out streams of invisible particlelike signals (words, gestures, and so on)? Sometimes forces seem to be like fields, part of the geometry of a situation, the tension between countries or within families, the dark cloud that seems to hover over some people, while a sunny streak lights the path of others. A field is like a perfume, a pervading influence. Some people seem remarkably good at sensing them. They know when the "vibes" are right, or when there is inherent tension in a situation.

But even a perfume can be broken down into "quantized" molecules that spread out from the person wearing it. If you throw out a compliment or a threat, you exert a force on someone. Sometimes it has the nature of a discrete event, or particle. At other times, it seems an integral part of the situation. Sometimes, you seem to "fall into" a good job or a "trap" just as the earth falls into its fortunate (for us) orbit around the sun. At other times, you can trace the job back to the letters you sent, the interviews you had, and so on.

Of course, in one sense, it doesn't matter a whit whether the force seems to act like a field or like a particle, whether it seems a discrete event or a property of space. All this is imagery on which different kinds of thinking can be based. What matters is what actually happens. Lo and behold, even in physics much of the terminology of forces has been altered since the introduction of quantum mechanics. Forces between particles are described much more accurately as "interactions." When two particles interact with each other, they exchange energy and/or momentum. This is exactly what happens when you stub your toe. The energy from the egg you ate this morning (which came from the sun by way of the chicken that ate the corn that absorbed the sun and then laid the egg) is converted into the electrical energy of nerves which is converted into the kinetic (or motion) energy of your muscles. When your toe hits the doorsill, some of that energy is exchanged with the doorsill. The energy of your kick heats up the molecules in the doorsill, and the doorsill kicks back some of the energy in the form of pressure you feel as pain.

Forces are obviously very real and important things, but describing them in the metaphors of everyday life can leave us somewhat muddled.* A force particle no more pushes other particles around than charmed particles are charming. Strictly speaking, a

---

*See "Science as Metaphor."

force is a transfer of energy and momentum. Two objects interact and things are never quite the same for either of them again. What happens in between is mostly guesswork. Or at least, it is not describable. As Bertrand Russell put it, a force is something like a sunrise—a convenient way of explaining something. A force no more literally forces something to happen than the sun literally rises: "Electricity is not a thing, like St. Paul's Cathedral; it is a way in which things behave. When we have told how things behave when they are electrified, and under what circumstances they are electrified, we have told all there is to tell."

Forces, in other words, are really ways of describing the way things are *connected* to each other. Inertia, action/reaction, relative forces, and fundamental forces all relate one part of the universe to another part of the universe. And the way we describe them, it turns out, really *does* matter.

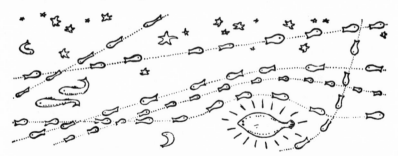

A flatfish named Albert saw that there was an unseen mound in the structure of space.

Sir Arthur Eddington points out the difference best in his story of two special fish, called "Isaac" and "Albert." Both Isaac and Albert were flatfish, living in a sea of two dimensions. Isaac looked at the motions of the other fish around him and noticed that all of them seemed to curve from their normally straight paths in a curious way. He discovered the "force" behind their curvature was another large fish, who attracted all the other fishes to him. As Guy Murchie puts it:

> This adequately accounted for most of the peculiar curves, so nobody bothered about the lesser attractions of a small moonfish circling nearby or the great numbers of fixed star-

fish twinkling in the background. And the only discontent-
ment left was the carping of a few carp who did not see how
the sunfish could exert such great influence from such a distance, though they presumed his influence must spread
forth somehow through the water.

Then along came Albert, who said that the fish were not attracted to the sunfish by *forces* so much as they were forced to
swim around the sunfish in curves because the sunfish was sitting
(or swimming, as the case may be) on top of a large mound. Of
course, the fishes could not directly sense such a mound because a
mound has three dimensions and the fish were two-dimensional.
In the same way, we three-dimensional creatures are unable to
sense the curvature of our own space which includes the fourth
dimension of time. And in the same way, it is the geometry of
space-time—the "mounds" in it caused by the presence of mass—
that result in what we have previously described as "forces" like
gravity.

It turns out that you can actually calculate and measure the
slight differences that would result depending on whether Isaac
was right, or whether Albert was right. And in several major
"tests," Albert's theories have been confirmed. So it does matter
whether the various parts of nature are knit together through
fields, forces, particles, and/or geometry. Or as the real Isaac
wrote in his *Principia* back in 1686:

I am induced by many reasons to suspect that [all the phenomena of Nature] may depend upon certain forces by
which the particles of bodies, by some causes hitherto unknown, are either mutually impelled towards one another,
and cohere in regular figures, or are repelled and recede
from one another. These forces being unknown, philosophers have hitherto attempted the search of Nature in vain;
but I hope the principles here laid down will afford some
light either to this or some truer method of [natural]
philosophy.

4. QUANTUM LEAPS

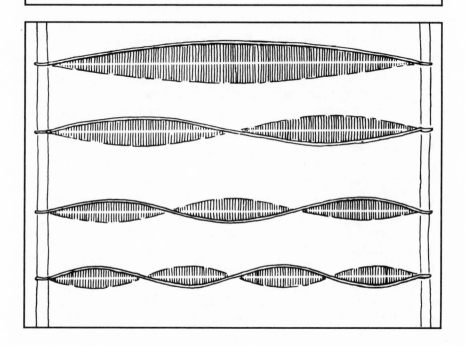

The greater part of this book is devoted to debunking the notion that nature's laws do not apply to human nature, that science is unfathomable, esoteric, and totally unconnected to people's everyday lives. I run into a problem, however, when I enter the realm of quantum theory. Almost inevitably, things quantum mechanical are characterized as weird, unseemly, and bizarre. Quantum mechanics comes on as so off-the-wall that only a mystic state of mind can even begin to probe its mysteries. Some popular interpretations get lost in the ambiguities of the notorious Heisenberg "uncertainty" principle (see below)—and come to the conclusion that the universe is only an unruly swarm of random events. Other accounts make much of the subjective nature of reality when read under the light of quantum theory. (Does a tree falling in a forest make a sound if no one is around to hear it? Does an electron really exist if no one can measure its position and motion at the same time?)

But there's nothing innately peculiar or surprising in the finding that we inevitably influence what we closely observe, or that perception is an active process, or that there are many substantial and important things in this world (beauty, for one) that are hard to put a physical finger on. Not that the nether world of the atom isn't brimming with weird and wonderful things; just that it is not necessarily any more weird or wonderful than gravity, or stars, or life. I tend to agree with Weisskopf when he says: "The only weird thing about quantum states is that they can't be described in ordinary language."

It is true, of course, that the introduction of quantum theory in the early 1920s marked perhaps the greatest revolution in all of physical science. It may have been the *only* revolution, for that matter. Weisskopf and others point out that most of the other giant steps in our understanding of nature were really *evolutionary*, in that they sprang from previously established foundations: facts were reorganized, or connected in new ways, or seen in a different context. Quantum theory, however, broke away completely from those foundations; it dove right off the end. It could not (cannot) adequately be described in metaphors borrowed from our previous view of reality because many of those metaphors no longer apply. But the net result has not been to obscure reality or make the nature of things more elusive and murky. On the contrary, most physicists would agree that what quantum theory has brought to science is exactly the opposite—concreteness and clarity.

WHAT IS QUANTUM MECHANICS?

Essentially, quantum mechanics is the mechanics of quantized things. What does that mean? Well, mechanics is the explanation of the way things work in terms of energy and forces and motion. Before the turn of the century, the way things worked was pretty well explained by Newtonian mechanics, also called "classical" mechanics. Everyone knows about typical Newtonian classical systems: billiard balls colliding and veering off on precisely predetermined paths, with precisely measurable amounts of energy and momemtum; planetary systems orbiting central suns according to the laws of gravity. Once Lord Rutherford had "seen" the atomic nucleus, the atom itself was considered just such a planetary system: electrons orbited the nucleus like planets orbiting the sun. And atomic particles behaved, of course, like billiard balls—albeit very *small* billiard balls. But the central feature of Newtonian mechanics was that everything was continuous; things flowed smoothly through space; energy could come in an infinite range of amounts; light undulated in continuous waves; there was no minimum amount of anything.

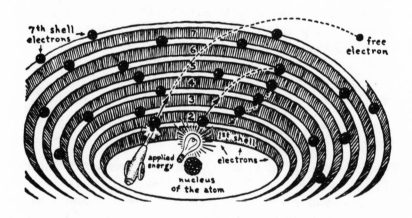

Guy Murchie's whimsical atom illustrates the nature of quantum leaps: the cannon balls, like electrons, need a minimum amount of energy to jump to the next level or "state." They cannot occupy a space "between" the levels.

Quantum mechanics changed all that. Now energy, light, force, and motion (among other things) are known to be *quantized*. You cannot have just any old amount; you can only have multiples of certain minimum quantities. Quantum mechanics meant that all the *qualities* of subatomic things—and by extension, of all things—were precisely *quantifiable*. In a sense, this has made things nice and neat and even more "scientific" in the popular sense. But it also meant that the smooth continuity of the Newtonian universe was lost. Now nature is revealed to be somewhat jerky or grainy, jumping from one quantum amount to the other, never traversing the area in between. Moreover, this leads to an uneasy uncertainty about what is going on *between* those quantum states, *during* those quantum leaps. In fact, it turns out that there is no way of knowing the precise state of things between quantum states. In human terms at least, there *is* no transition between quantum states. You can have either one or two or three units of energy or momentum or light or force or matter or whatever, but there is *no such thing* as one and one-half or two and three-quarters units. Everything in the quantum mechanical universe (which is our universe, of course) happens in quantum leaps. The exact nature of the so-called uncertainty associated with these leaps is discussed at length in the chapters that follow—especially "The Measured Approach" and "Natural Complements." But in any event, these uncertainties remain relatively unimportant, as Weisskopf always says, until you actually set out to *measure* things as small as atoms (hardly an everyday affair) and try to describe them in the language of typical Newtonian systems.

THE DEFINITENESS PRINCIPLE

Weisskopf goes so far as to say that the Uncertainty Principle should really be called the Definiteness Principle. His reasons are compelling: consider a typical Newtonian (pre-quantum mechanical) system—nine planets orbiting a central sun. The laws of "classical" physics allowed for a tremendous amount of flexibility. All the law of gravity requires is that the nine planets orbit the sun in more or less elliptical paths; nothing says that the orbit of the earth or of any other planet must be "just so." In fact, depending upon the initial conditions present at the formation of the solar system, almost any elliptical orbit would do. And if someday a passing star should come along and push our world out of its place in the sun with the force of its gravity, our orbit would be irrevocably changed. There is nothing whatever special about our present position that would cause us to remain here—or to spring back to our original orbit once we had been pushed out of it. Since planets

The energy patterns of any particular atom are exactly the same whether the atom comes from within your body or from a far-off star. The quantum states of the hydrogen atom shown in this picture are models; if you really tried to photograph them, they would be destroyed.

orbiting a star can fall into almost any possible elliptical path, other solar systems surrounding other stars are likely to be found (if they are found at all) in a great variety of arrangements.

Contrast this with an atomic system—which was pictured as just such a miniature solar system. The central nucleus attracts the orbiting electrons just as the sun attracts the orbiting planets. But here the similarity ends. In solar systems, an infinite number of stable configurations is possible; in atomic systems, only about a hundred—each corresponding to one of the known elements. If the system contains one nucleus (one sun) and one electron (one planet), it takes a single, unchanging form—the hydrogen atom. All hydrogen atoms are exactly alike. And if a bunch of hydrogen atoms get bounced around by other hydrogen atoms or by anything else, they automatically spring back to their original shape.

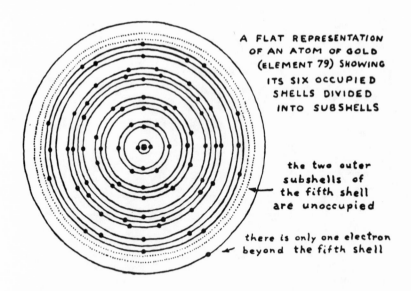

A FLAT REPRESENTATION
OF AN ATOM OF GOLD
(ELEMENT 79) SHOWING
ITS SIX OCCUPIED
SHELLS DIVIDED
INTO SUBSHELLS

the two outer
subshells of
the fifth shell
are unoccupied

there is only one electron
beyond the fifth shell

Every gold atom in the universe is exactly alike.

"Before we got quantum theory," says Weisskopf, "our understanding of nature did not correspond at all to one of the most obvious characteristics of nature, namely, the definite and specific properties of things. Steam is always steam, wherever you find it.

Rock is always rock. Air is always air. Two pieces of gold mined at two different locations cannot be distinguished." Even two pieces of gold found in different *galaxies* would have identical properties.

It should be highly improbable to find two atoms exactly alike—just as it would be highly improbable to find two solar systems exactly alike. But obviously, it's not. A carbon atom enters your body as part of a sandwich and emerges hours or days or perhaps even years later after taking part in countless chemical reactions still as a carbon atom. The particular combination of protons, neutrons, and electrons that makes up a carbon atom can *only* be arranged "just so." Quantum theory brought just this exactness into our understanding of atoms. If anything was uncertain, it was *classical* systems.

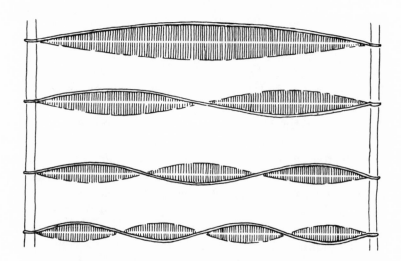

A string attached at two ends can vibrate only so that one, two, three, four, etc., half wavelengths fit in the space between the attachments. You could say that the string has a strictly limited number of "allowed states"—just like the quantum mechanical atom.

Too many good books have already been written on quantum theory for me to even attempt to go into the details of its development in this short chapter. Suffice it to say that it was Danish physicist Niels Bohr* in the early 1920s who came up with a way to understand the stability and exactness of atoms using the analogy of standing waves. Take a jump rope, or a violin string, secured at both ends. If you pump energy into it and set it swinging, it can vibrate only in a certain number of ways, taking a few predetermined shapes. A violin string vibrates in its fundamental frequency, or in twice, three times, or four times that frequency—in other words, in its characteristic harmonics. It cannot vibrate in two and one-half times that frequency. If you imagine an electron "wave"† confined to an atom in much the same way, you can easily understand how it would be forced to assume only a certain number of predetermined vibrational "states." Atoms—unlike solar systems—are innately stable and consistent because electrons cannot take their places around the nucleus just anywhere. Every hydrogen atom in the universe strikes the same chord of frequencies. (Weisskopf told me he once tried to play the chord for hydrogen on his piano: "It sounds terrible," he said, "but then it's not music for our ears.")

There are other analogies, of course, though none quite as useful as the image of standing waves, since electrons in fact do behave like waves. But just for argument's sake, one could imagine the electron in the atom as a child jumping up stairs. Depending on the amount of energy she has, she can jump up to the second stair, or the fourth, or the sixth, but she cannot land safely or remain stable at step two and one half, or three and one quarter. She needs a minimum amount of energy before she can attain the next step, or state. If she doesn't have quite enough energy to make it to step four, she will remain at step three. And atoms will not absorb radiation unless the energy they receive contains the minimum to make the next "quantum leap."

Or you could think of the girl as a marble trying to jump into a box. It needs a minimum amount of energy to get into the box. Of

---

*Of course, the development of quantum theory did not take place in quantum leaps, but happened over a period of many years and involved many different physicists.

†One of the most intriguing discoveries of twentieth-century physics was the finding that all particles sometimes behave like waves—and all waves sometimes behave like particles. See "Natural Complements."

course, it also needs a minimum amount of energy to get *out* of the box. And a child jumping *down* stairs also behaves somewhat like an electron changing states within an atom. In this case, she gives off energy to the floor as she jumps to a lower state. But the energy comes in clumps. A jump from step four to step two gives off two "steps'" worth of energy, and so on.

Niels Bohr (left) *and Max Planck, and quantum transitions in a hydrogen atom.*

An electron jumping to a lower state gives off its energy in the form of light. A jump from "orbit" or step five to orbit three might radiate clumps of red light; a jump from six to two might radiate more energy (higher frequency) blue light; a jump from orbit two to orbit one (the ground state) might give off very low energy radio waves; a jump from orbit eight to the ground state might give off a highly energetic X ray. There is a similar series of even more energetic quantum states within the nucleus that account for radiation such as gamma rays.

These quantum leaps are clearly visible in many everyday phenomena (so one can certainly see where quantum theory helped to clarify otherwise unfathomable things). Take the spectra of common elements, for example. If you look through a prism at neon gas or mercury gas or sodium gas of the kind common in lamps, you will see a series of sharp colored lines. Each line represents a quantum leap from one energy state to another. And the spectrum of sodium or mercury or hydrogen looks exactly the same no matter where it's found—even on the farthest stars.* If you've ever wondered how scientists seem to know what distant stars are made of, this is behind the technique they use. In fact, these unmistakable atomic fingerprints are even used to determine the chemical makeup of clues left at the scene of crimes.

Each quantum leap between energy levels in an atom corresponds to a specific frequency (or color) of light. This gives each atom a unique set of spectral "fingerprints" that can be used to identify it even in a remote star.

Even the fluorescent lamps in your home radiate light in discrete quantum jumps, the colors depending on the gas inside the tube—usually mercury vapor. However, the insides of the tubes are usually coated with phosphors that reradiate the light in a range of other colors designed to mimic the light that comes from hot metals like the filaments in incandescent lamps. So if you look at them through a prism, you will probably see the sharp lines of mercury somewhat obscured by the continuous spectrum of the phosphors.

*Actually, the spectrum doesn't look *exactly* the same on stars, but the differences tell astronomers things such as how fast the star is moving and how far away it is.

Hot metals glow with a continuous spectrum of colors, and not with the sharp spectral lines characteristic of quantum leaps, because they emit light in a different way from gases. George Gamow had described the spectrum of an excited gas as the harmonics of a single instrument, but atoms in a solid are so closely packed together that it is as if you threw all the instruments of a symphony orchestra into a bag and shook them up. The spectral colors that shine from hot metals come not from electrons bound to atoms, but from free electrons that are, well, free to vibrate more or less at random.* They do not have to jump from step four to one, or six to two. They can give off a whole range of frequencies of light. For this reason, anything that gets hot enough burns with the same colors; it does matter what is on fire.

(Sometimes, so much energy can be pumped into an atom that the electrons dislodge entirely. It ceases to be an ordinary atom of carbon or neon or oxygen and instead becomes an ion—an electronically charged "part" of an atom. If *all* the electrons are knocked off, then you are left with a special kind of gas known as a plasma—an amorphous mixture of nuclei and electrons. There are no atomic quantum states in plasmas [there are no *atoms* in plasmas]. Yet most of the matter in the universe exists in this highly energetic form. Stars are essentially balls of plasma; the spectral lines that tell what stars are made of come from the cooler atoms on the star's surfaces.)

At earthly temperatures, however, the quantum reigns. "Ultimately," says Weisskopf, "all the regularities of form and structure that we see in nature, ranging from the hexagonal shape of a snowflake to the intricate symmetries of living forms in flowers and animals, are based upon the symmetries of these atomic patterns." The fact that you inherit stable genes from your parents is

---

*The frequency of vibration of light—that is, how rapidly the light waves oscillate or move "up and down"—determines its color. Higher frequencies produce shorter waves; in fact, if you shake or vibrate a rope, you can see that faster vibrations make shorter waves. It also takes more *energy* to make smaller, more rapidly vibrating waves. With light, too, short waves (fast vibrations) correspond to high energies. In the electromagnetic spectrum, the most slowly vibrating "light" falls into what we call the radio frequencies; these waves can be as large as mountains. Visible light waves average about 1/50,000 of an inch long. X rays are as small as atoms.

based on the inherent stability of the quantum states of the molecules that make up DNA. The hardness of the metal that makes up my typewriter, the softness of my baby's skin, the scent of flowers wafting in through the window—all are consequences of the quantum states of atoms. Tell that to the next person who thinks that quantum mechanics is something farfetched and far removed from the stuff of everyday life!

One thing that really seems to bother people about the idea of quantum states is how they can mysteriously materialize out of nowhere—how can you get from A to B without trespassing on the territory in between? This quantum leaping in and out of existence is as unnerving as the elusive appearances and disappearances of the smile on the Cheshire cat. A flight of stairs may seem discontinuous, but at least the steps (like islands) are connected underneath—and a child jumping up or down them passes through the air in a continuous trajectory before landing on one or the other.

An electron, however, cannot exist between quantum states—not even for an instant. There is no such thing as "in between." No wonder this offends people's philosophical sensibilities: it is like jumping from one hour to the next without passing through the minutes in between, or disappearing from the universe at one end of a room only to reappear miraculously at the other. Now you see it, now you don't.

I suppose it would be bad enough if only the permissible energy states of atoms were quantized—even if that means that the light waves they radiate are necessarily quantized and particlelike, too. But it turns out that virtually everything in the subatomic world is quantized. Not only energy and light, but also matter, "action" (energy × time), momentum, spin, electric charge, and all the other exotic qualities of subatomic things—like "strangeness" and "charm." You cannot even conceive of an action, or a motion, or a bit of matter, smaller than this minimum. An atom absorbing any one of these qualities has to swallow it whole, or not at all; it spits it out in quantum clumps. This means that the very stuff of the universe cannot be smoothed out beyond a certain point; it has a texture; it is grainy, or lumpy.

In truth, this seems odd and unpleasant only until you stop to think about it. After all, many things in everyday life are lumpy, too—auto transmissions, for example. As Gamow points out, the gearbox on your car is something like the quantum states of atoms. "One can put it in low, in second, or in top gear, but not in between." (Of course, you *can* physically put it in between, but it doesn't have much meaning as far as the motions of your car are concerned.) Another familiar phenomenon that comes in clumps is people. Richard Feynman uses the familiar example of the "average American family" consisting of 2 adults and 2.2 children, or whatever. Everyone knows it's a joke because everyone knows that children, like quanta, are units. I would even argue that states of mind—even leaps of the imagination—are similarly quantized. There is rarely a continuous transition from one idea or state to the next. (As Einstein said, "There comes a point where the mind takes a leap—call it intuition or what you will—and comes out on a higher plane of knowledge.") Cultures, perceptions, beliefs, and even phases of life often seem as discretely separated as individual quantum states—which is why we sometimes feel truly transformed when we move from one to another. This is not entirely surprising because, of course, even the brain is quantized: an impulse is transmitted by a nerve cell as a whole or not at all.

Most of this is metaphor, of course. But the metaphors come easily enough to suggest that a quantum is not a completely unfamiliar concept. In essence, it means only that some things need to be considered whole: children, snowflakes, atomic states, and also memories, experiences, poems, paintings, and a whole host of other things. It also embodies an irreducible on/off, yes/no quality—something that should be intimately familiar to anyone weaned in the computer age. A person, for example, can have a whole range of degrees of a quality such as charm, but when it comes to atoms, either you've got it or you don't. The same is true of energy, spin, "strangeness," electric charge, and so on.

On a less metaphorical level, almost anything that has to do with waves has these very specific properties reminiscent of quantum states. Not only violin strings, but also the air inside flutes and organ pipes vibrates in fundamental frequencies or exact multiples thereof—each time taking a quantum leap to the next harmonic. Even a single wave is a quantum in that it has to be considered whole. It makes no more sense to talk about a third of a wave than it does to talk about a third of a child.

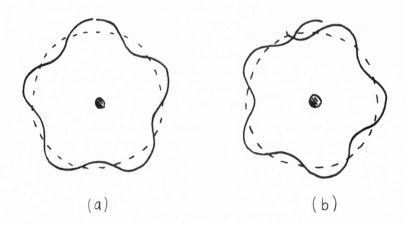

(a)   (b)

The electron orbits in an atom are very specific because the circumferences of the orbits are exact multiples of the electron wavelengths. (a) is a stable state. (b) is not "allowed."

In another context, Stephen Jay Gould argues that our fear of the quantum leaping comes from "a deeply rooted bias of Western thought that predisposes us to look for continuity and gradual change: *natura non facit saltum* (nature does not make leaps), as the older naturalists proclaimed." Gould is one of a new breed of naturalists who are convinced that even evolution probably occurred in something similar to quantum leaps. "Change is more often a rapid transition between stable states than a continuous transformation at slow and steady rates," he writes, in terms that could also apply to atoms. Evidence accumulated from the fossil record seems to deny older theories that evolution proceeded gradually through small continuous adaptations; rather, species seem to appear, stick around for a while virtually unchanged, then disappear again. Gould was surprised to find that many Soviet scientists already shared this perception—at least in part, he speculates, because of their training in the dialectical laws, laws that suggest "that change occurs in large leaps following a slow accumulation of stresses that a system resists until it reaches the breaking point. Heat water, and it eventually boils. Oppress the workers more and more and bring on the revolution."

QUALITY AND QUANTITY

This notion that merely having *more* of something (or less of it) can change the nature of things *qualitatively* is at the heart of quantum mechanics. But it also can be one of the hardest aspects to accept. Physicists tend to give people the creeps when they start describing the fundamental building blocks of nature in terms of quantum "numbers." How can the addition or subtraction of single units of spin or electric charge or other "quantities" make the difference between neon and sodium? Or ultimately even apples and oranges? The immediate explanation, of course, is that units of electric charge, for example, determine the chemical properties of things—but there is a much more interesting and fundamental relationship.

The first time I spoke with M.I.T.'s Philip Morrison about this, he offered what I thought was a very good example: "Dust, sand, pebbles, rocks, and boulders are all made of the same material. They just contain different quantities of the same stuff. But we view them as qualitatively different. So in that sense, quantity *is* quality."

One of the first people to recognize the intimate connection between quality and quantity was the sixth-century B.C. Greek philosopher Pythagoras. He discovered that the pitch of a note depends on the length of the string producing it, and that pleasing intervals are those that correspond to simple mathematical ratios: 2:1 producing an octave; 3:2, a fifth; 4:3, a fourth; and so on. The quality of musical tones—like the qualities of elements—is based on the numerical relations of their parts.

Today, the evidence that quantity effects quality is everywhere. The quantity of money or education you have may or may not make a difference in the quality of your life, but the quantity of pollution, noise, and crime in your neighborhood almost certainly does. Rapid increases in the overall quantity of otherwise beneficial things (cars, plastics, houses—even people) can turn them into qualitative disasters. Medicine taken in too large doses can be poisonous. And bigger (nuclear) bombs do not merely produce bigger wars: the difference between killing millions of people and wiping out civilization is qualitative, not quantitative.

One of the most impressive qualitative changes produced by sheer increase in quantity is the human brain. "When people evolved from the animal kingdom," says physicist Weisskopf, "something new must have happened. We contend that this new element is based solely upon a quantitative difference in the nervous system. By an increase in this system, nature established a new type of evolution that broke, and will break, all rules established in the previous evolutionary periods." Naturalist Gould is similarly impressed: "Perhaps the most amazing thing of all is a general property of complex systems, our brain prominent among them—their capacity to translate merely quantitative changes in structure into wondrously different qualities of function."

It takes a large quantity of atoms to make an organic molecule, and a large quantity of organic molecules to make a person, and a large quantity of people to make a crowd or a country. But in each case, it's clear that the difference is a lot more than quantitative.

Weisskopf talks a great deal about quality and quantity (in quantum mechanics and other things), so I will feel free to borrow one of his favorite familiar examples. Consider a container of water standing around at room temperature. Molecules of water are continually evaporating and entering the air as water vapor; meanwhile, a more or less equal number of water molecules are condensing out of the air and entering the container of water. Now say you start to increase the quantity of heat in the water—raising its temperature. At first, you see the expected quantitative changes: more water molecules leave the surface than before. But pretty soon you see a dramatic *qualitative* change. When the temperature gets hot enough, the water disappears! The boiling "destroys" the water, in Weisskopf's words. But in fact, nothing special has happened—only a little more of what happened before.

This ability of quantitative changes in temperature to produce dramatic qualitative differences in matter goes to dramatic extremes at extreme temperatures. As mentioned before, matter at very high temperatures actually falls apart; electrons leave the atomic nuclei and form an undifferentiated plasma, the stuff of stars. Plasma is nothing like ordinary earthly matter. For one thing, its electrically charged particles are not bound together in atoms, but behave independently. For another, it does not come in different well-defined varieties—such as silicon, oxygen, and lead. At even higher temperatures, an even more extreme state of matter is created in which even atomic nuclei come apart into their constituent protons and neutrons. At higher temperatures still (perhaps at the origins of the universe), even protons and neutrons disintegrate into a whole host of exotic species. If the esoteric anti-particles and mesons and J/psi's that pop up in high-energy accelerators seem strange to us, it is because they represent a *qualitatively* different kind of matter than we are used to.

Boiling destroys water.

As temperatures go down, matter again transforms into a whole series of qualitatively different states. The transition from gas to liquid takes place when a certain quantitative difference in temperature is reached; solids and crystals form at lower temperatures still. At what are known as "supercold" temperatures (near absolute zero—about minus 450 degrees Fahrenheit), there exists yet another state of matter—equally exotic as its superhot counterparts. Certain metals at supercold temperatures become "superconducting": they can carry an electric current forever with resistance. Supercold helium becomes a "superfluid" that flows up and out of bottles, and down through the bottoms of ceramic containers. Matter becomes so ordered and so quiet that subtle quan-

tum effects make themselves visible. When all the random motion
that makes up heat is taken away, one can hear the inner whispers
of atoms "like the murmurings of a seashell," as one physicist
put it.

But temperature isn't the only physical quantity that makes
astounding qualitative differences in the way things behave. Take
speed, for example. All the amazing effects of special relativity—
from time dilation to the eerie increase in mass of fast-moving objects—only come into play when the velocity is quantitatively
close to that of light.* These things do happen at lesser velocities,
of course. But they are too insignificant to make a qualitative difference. Another prime example is quantity itself: a single atom or
a single coin or a single person behaves in a qualitatively different
way from a trillion atoms or a hundred coins or a thousand people.
For one thing, the behavior of the latter grouping is somewhat
predictable; the behavior of the former is not.†

Size may be the most significant example of all. A quantitative
difference in size alone is enough to make a mockery of most of our
everyday, human-scale models. A billiard ball the size of a star
does not behave like a billiard ball the size of an atom. A ball of
matter the size of an ordinary billiard ball cannot collapse under its
own weight, but a ball of matter the size of a star easily can. A ball
of matter the size of an atom doesn't behave like a ball of matter at
all, but more like a wave. Curved space is primarily a property of
huge things; quantum effects apply only to tiny things. On the
scale of an atom, gravity is insignificant; on the scale of the universe, it is the most significant force there is. Indeed, one of the
most pressing preoccupations of present-day physics is to find the
connections between the behavior of the very large, and the very
small—the so-called quantization of gravity.

Sir James Jeans pointed out that while philosophers generally
think in terms of *qualities*, physicists normally describe things according to their *quantities:* "The philosophical lecturer may be telling his audience that a lump of sugar possesses the qualities of
hardness, whiteness, and sweetness, while his colleague in the science room next door may be explaining coefficients of rigidity, of

* See "Relatively Speaking."
† See "Cause and Effect."

reflection of light and hydrogen-ion concentration—measures of the degree to which the qualities of hardness, whiteness, and sweetness are possessed."

And yet it is difficult to draw a clear difference between the two. For if hydrogen-ion concentration is a measure of sweetness, and reflection of light determines whiteness, then quality *is* quantity in a very real sense.

For that matter, talking about quantum mechanics at all involves taking a kind of quantum leap into a new dimension where almost everything is qualitatively different—and mainly because the things we are talking about are quantitatively so much smaller. Einstein knew that going to extremes of size or speed could lead to qualitatively surprising results. Throughout the history of science, the most fruitful areas for exploration have hovered at the extremes and fringes—the outer limits of hot and cold, fast and slow, big and small, few and many. Or as a physicist friend once told me: "Everything in the middle is engineering."

So if quantum mechanics seems weird to us, perhaps it's only natural. Atoms—from our perspective—are unimaginably small. It should not be surprising that they do not behave as we do.

5. RELATIVELY SPEAKING

Almost from the beginning, relativity has been considered as much a philosophy as a physical theory. Einstein himself guessed that more clergymen were interested in it than physicists. Perhaps this had to do with the fact that relativity touches deep roots in culture, history, and religion: our world view of things depends very much on cosmic notions of space and time, which in turn determine how people fit into the scheme of things. Yet whatever the reasons, attempts to popularize and philosophize about relativity have almost always got it wrong. What's filtered down as the popular perception of relativity is the phrase "Everything is relative." In fact, the implications of Einstein's ideas are almost exactly the opposite.

What many people mistake for the theory of relativity goes back at least as far as Galileo, and it has to do primarily with relative motion. That is, if you were riding inside the closed cabin of a steadily moving ship, you could not tell whether you were truly moving or not.* You could perform all manner of experiments—throw balls, swing pendulums, try to balance on your toes—and everything in the cabin would behave exactly the same whether you were moving or standing still. The reason is that while you would be moving relative to the world outside the ship, you would not be moving relative to the world inside it—just as when you are relaxing on your door stoop you are not moving relative to the rapidly spinning earth, and thus you perceive no motion. Motion is relative because it depends on your point of view. What's moving in one frame of reference is not necessarily moving in another.

Say you asked a person standing on a nearby dock to tell you whether or not you were moving. The person could answer: "Yes, of course you are moving, you dummy, can't you see the dock moving past you?" To which you could reply: "But how do I know that it's I and not the dock that's moving?" (And of course, the dock *is* moving along with the motion of the earth, but that is another matter.) Or else, the person on the dock could answer: "You are moving relative to my dock, but as long as you stand still in the cabin there, you are not moving relative to your ship." Or else, the person could shrug her shoulders and say: "Well, everything is relative, you know."

The question would be much more difficult, of course, if you wanted to find out whether or not the *earth* was moving by asking such an outside observer. For where would you put her "dock"? How could you find her a frame of reference where she would be standing still? The earth moves relative to the sun, but the whole solar system moves relative to the galaxy, and the galaxy moves relative to the rest of the universe, and by all indications the universe is moving (in some sense) too. But relative to what?

"One of the great clichés about Einstein's theory is that it shows that everything is relative," wrote James R. Newman in

*See "Seeing Things."

Science and Sensibility. "The statement that everything is relative is as meaningful as the statement that everything is bigger. . . . If everything were relative, there would be nothing for it to be relative to."

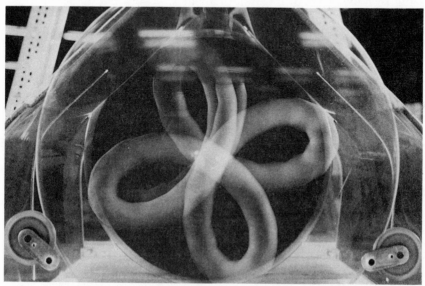

This photograph shows the pattern made by a ball moving back and forth in a straight line on a rotating turntable. If you moved along with the turntable in the same way as you move along with the rotating earth, you would not see this pattern; you would see the ball move in a straight line. Motion is relative because it depends on your point of view.

THE THEORY OF ABSOLUTISM

Einstein took Galileo's relativity and turned it on its head so that it came out what Weisskopf likes to call the theory of absolutism. Einstein looked inside the cabin of the moving ship (except it was a light beam), and was profoundly impressed *not* that everything is relative, but that the laws of nature stay exactly the same. Jump up, and gravity pulls you down the same way whether you are moving or standing still. Throw a ball, and it follows the same path. Water flows, clocks tick, raindrops fall, and electricity attracts the same whether you are moving or at rest. "Relativity established that the laws of nature are absolute and do not depend on the motion of the system," says Weisskopf. "It's *because* they are absolute that you cannot tell whether you are moving or at rest."

Of course, certain things that people *thought* were absolute have to be sacrificed in order to achieve absolutism in the laws of nature. Space and time become relative, but in the scheme of things space and time turn out not to be very important, or at least not as fundamental as many other things.

What is much more important is that the conventional concepts of physics embodied in Newton's laws simply don't work at very high speeds or under conditions of extreme gravity or in many other situations. Newton's laws would not hold true in all frames of reference, so the laws of nature would depend on whether you were moving and what system you were in. The laws of nature would depend on your point of view. Now that would be a theory of relativity!

Galileo and Newton realized that motion was relative, but insisted that space and time were absolute. Einstein saw that space and time—and energy and mass, for that matter—were also relative, but that all this was but a peripheral effect of the absolute nature of other fundamental constants—among them the speed of light. For all the odd effects of relativity flow from the even odder fact that the speed of light is always absolute—an absolutely unchanging 186,000 miles per second from any point of view, from any frame of reference, moving or not. The only reason "everything is relative" is that the speed of light and the laws of nature are not. And since light itself is nothing but the motion of electric and magnetic fields relative to each other—forces which are behind everything from the nature of matter to all human processes including perception—Einstein clearly hit upon a very fundamental "absolute" frame in which to construct his relative universe.

THE RELATIVITY OF TIME

Before getting to absolutes, however, it's best to test the waters of relativity by sticking some tentative toes in something easy and familiar—the relativity of time. In fact, it turns out that the relative part of relativity is not that hard to get used to: the idea that things can be relative is a common experience of everyday life. Absolutism, on the other hand, is much more difficult; therefore, it will wait until later. But in any case, there's little point in this short chapter of trying to cover the same ground that is covered so well in the many good popular books on relativity. Better to concentrate on what seems to be the heart of the matter—something that often gets lost in the litany of bizarre relativistic effects that include everything from time dilation to black holes. And the heart is simply this: that certain things are relative and certain things are absolute, and which are which is important.

Newton wrote in his *Principia:* "Absolute space, in its own nature, without relation to anything external, remains always similar and immovable. Absolute, true, and mathematical time, in itself, and from its own nature, flows equably without relation to anything external."

This concept of time may seem natural enough at a quick glance. Almost everyone accepts it unthinkingly. But in truth, it would make sense only if your notion of God is a great universal clock. Because every idea of time that you can possibly think of is intimately connected to a concrete physical event—the swing of a pendulum, the orbit of the earth, the vibrations of a quartz crystal, the quantum leaping of atoms, the motions of magnetic and electric fields, the lives of suns, and so on. Without such events, what would time consist of ? You can't have time in a void because there would be nothing, so to speak, to "tick." Time makes sense only when it's connected to things.

"Imagine a billiard ball as the only inhabitant of the universe," writes physicist B. K. Ridley.

What position does it have? The question has no meaning, for position can only be defined with respect to another position. . . . What size is it? But we have nothing to compare it with, so what possible answer can there be? Surely it endures. Since it does not change; since nothing can happen, nothing coming along and bouncing off it, how can we tell the passage of time? We cannot. Both Space and Time have no meaning whatsoever. So much for absolutes.

All standards of time are derived from the relationships between events that take place in the physical world. The time interval we call a year marks a single revolution of the earth around the sun; the day is a single spin of the earth around its axis; and long before humankind was born, the month probably matched the orbit of the moon around the earth. It does so no longer simply because the moon's orbit is continually changing as the moon moves farther away. Clearly, none of these astronomical measures of time could ever be considered absolute if only because they *are* constantly changing. Some five hundred million years ago, our day was only twenty and a half hours long. And just to remove ourselves from our parochial earth perspective for a moment, consider the concept of "time" on a planet like Mercury, where the day is longer than the year. It's a difficult idea to get used to because we are so accustomed to thinking of "days" as the natural division of "years" into 365 parts. We forget that the day, like the year, is a happening—not an empty interval of some amorphous quantity we call time.

Other standards of time have no natural origins at all. What is the natural origin, for example, of the seven-day week? Some people say it derived from Genesis (on the seventh day, we rest). Others link it to the seven notes on the familiar Western musical scale which corresponded to the seven classical Pythagorean planes (sun, moon, and five visible planets), which in turn played the music of the spheres. Various cultures at various times have tried out eight-day weeks, five-day weeks, and even a decimal ten-day week.

The hour is actually a quite recent addition. Until the fourteenth century, days were divided into much less regular intervals of morningtide, noontide, and eventide. The first hours were flexible: they varied from summer to winter, and from daylight to darkness. Well up into the Middle Ages, each day (dawn to dusk) and night (dusk to dawn) was divided into twelve equal parts. This meant that the hours of a summer day lasted much longer than the hours of that same summer night. Winter daylight hours were correspondingly shorter, and winter night hours correspondingly longer. Even regular hours would not do, however, when the Industrial Revolution made it necessary for the trains to run on time and the workers to arrive for the five o'clock shift, so minute hands sprouted on clocks like so many offspring of a new age.

But all these concepts of time are modern inventions. As Arthur Koestler pointed out, less than fifteen generations ago "time . . . was simply the duration of an event. Nobody in his senses would have said that things move *through* or *in* space or time—how can a thing move in or through an attribute of itself, how can the concrete move through the abstract?" The idea that time was an attribute of things—like length, width, and depth— seems much closer to Einstein's four-dimensional space-time than the abstract clock time that we carry around with us today.

Most of all, perhaps, time is a form of perception. It is certainly accurate to speak of our "sense" of time. "Just as there is no such thing as color without an eye to discern it, so an instant or hour or a day is nothing without an event to mark it," writes Lincoln Barnett. We go about marking it and discerning it in an astounding variety of ways. Setting clocks aside for the moment, everybody knows that perceptions of time change dramatically from country to country, from person to person, and even within the same person, well, from time to time. "A watched pot never boils" is a statement about the relativity of time."Time flies when you're having fun" is another. Because we measure time by the events that mark it, it should not be surprising that our sense of time is profoundly influenced by the nature of the events themselves.

Time also changes with perspective. I am reminded of the time I was talking with M.I.T.'s Philip Morrison about the extremely short lifetimes of subatomic particles and he said: "Short compared to what? Short compared to *our* lifetimes, you mean." The atomic year of hydrogen, at 10^{-16} seconds (one divided by the number ten with sixteen zeros after it) is infinitely short compared to an earth year. (An atomic year is the time it would take an electron to "orbit" the nucleus if the atom were a miniature solar system.) Yet the atomic year itself is infinitely long compared to the fundamental time units pertaining to nuclear particles, which are millions of times shorter. To a particle in the nucleus, an electron would seem almost stationary—just as the "fixed" stars seem stationary to us.

There is a wonderful passage in Tracy Kidder's book *Soul of a*

New Machine about time inside a computer: "It's funny, I feel very comfortable talking in nanoseconds," says one of the computer engineers, speaking of the term for *billionths* of a second.

> I sit at one of these analyzers and nanoseconds are *wide*. I mean, you can see them go by. "Jesus," I say, "that signal takes twelve nanoseconds to get from there to there." Those are real big things to me when I'm building a computer. Yet when I think about it, how much longer it takes to snap your fingers, I've lost track of what a nanosecond really means. Time in a computer is an interesting concept.

In geological time, on the other hand, an "instant" can be ten million years—because ten million years is but 1/450 of the earth's history. A thousand years is an interval so short that geologists can rarely resolve it. It's all but undetectable, and so is treated as a passing moment. "With our short memory," writes naturalist Loren Eiseley, "we accept the present climate as normal. It is as though a man with a huge volume of a thousand pages before him—in reality, the pages of earth time—should read the final sentence on the last page and pronounce it history."

Or as Guy Murchie writes in *Music of the Spheres:*

> A million-year-old mountain is just a shoulder-shrug to an earth that already counts her years in billions. Like matter itself, *terra firma* thus reveals her illusory nature. . . . Of such are mountains born and raised—too slowly for mortal eye to see—too slowly to be noticed by the full sweep of human history—yet literally bursting and buckling and boiling in the view of the patient moon, who has reason enough to know the meaning of violence as anyone can see in her pitted face.

The universe is atick with all kinds of time-keeping devices, all clocking different kinds of time. Radioactivity provides a natural clock that resides within atoms. That is, certain atoms exist in unstable forms that after a while decay into more stable forms. An ordinary carbon nucleus, for example, contains six protons and six neutrons and is known as carbon-12. But there is another form or "isotope" of carbon—carbon-14—whose nucleus contains six protons and *eight* neutrons. During radioactive "decay," one of those extra neutrons emits a negatively charged electron and becomes a positively charged proton. And faster than you can say "alchemy," the unstable carbon nucleus has changed into a stable nitrogen nucleus with seven protons and seven neutrons.

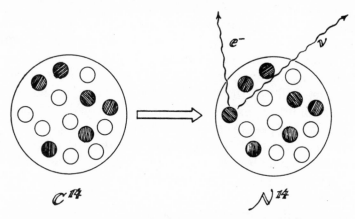

Radioactive transformation of C^{14} to N^{14}. Open circles are neutrons, dark circles protons. One neutron to the extreme left of C^{14} changes into a proton and emits a negative electron (e^-) and a neutrino (v).

What's even more fascinating about these transformations, however, is their *timing*. Every 5,700 years, exactly half of a given number of carbon-14 nuclei will decay into nitrogen-14 nuclei. If you started with ten trillion carbon-14 atoms, you would have five trillion carbon-14 atoms exactly 5,700 years later. And 5,700 years after that, you would have exactly 2.5 trillion carbon-14 atoms. It is as if you had a used car lot where exactly half of the cars decayed each year: if you started with one hundred cars, fifty would rust away the first year, twenty-five (or half of the remainder) the second year, twelve and a half (or half of that remainder) the third year, and so on. A year, then, would be the natural "half-life" of the cars on your lot, just as 5,700 years is the natural half-life of carbon-14 atoms.

The half-lives of atoms are remarkably accurate timekeepers because nothing outside the atom can affect them. They are immune to external influences. And unlike astronomical measures, they never change. Still, they provide a strange kind of clock that ticks only for large numbers of atoms. It does not work at all for a single atom. It is a clock based on the statistics of probability.*

And then, of course, there are all manner of biological clocks. All living things need to tell time to survive, to coordinate their internal functions with the clocks of the outside world, to know when to hibernate, or fly south, or sprout, or shed, or grow a winter coat. Hearts need to know when to pump, and lungs when to breathe. Different organs in the same body may keep different times to different kinds of clocks, releasing chemicals in concert with communications from a central brain. Stomachs, livers, sleep centers, may all tick to different, yet coordinated, tunes. The limits of our ability to perceive time intervals even determine how we see the world: if people could sense intervals shorter than 1/24 of a second (they can't), they would see the dark gaps between the frames of a movie; if they could perceive much longer intervals of time, on the other hand, they would actually be able to "see" plants (or children) grow.

Times change, of course, especially as we grow older. What adult isn't sometimes driven to distraction by a child's frenetic pace, and what child doesn't lose patience with his parent's slow-motion, hippopotamus approach to things? Many people have surmised that our quickening sense of time depends upon the diminishing percentage of a lifetime that each hour or year takes up as we age. To a year-old baby, a year is a lifetime—eternity. To a ten-year-old, a year is but a tenth of his lifetime, and each hour is proportionately shorter. "When he reaches fifty," writes Murchie,

> time is passing five times faster still, clocks have begun to whiz and a year is but 2 percent of his life. And if he reaches a hundred, it's 1 percent. His old friends have been dying off at a fearful rate while new children sprout into adults like spring flowers. Strange buildings pop up like mushrooms. A whole year to him actually consumes less conscious time than did four days when he was one year old.

Counting off intervals of time is a mysterious and highly indi-

*See "Cause and Effect."

vidual process. How does a plant know when to sprout? What kind of clock does a tree use to keep track of the passing years? How do our internal biological clocks know when to tick? Obviously, the answers have to do with some combinations of chemical reactions, and the ability to recognize patterns of warmth and cold, light and darkness.

But when it comes right down to it, it seems that people have come up with all manner of measures of time without necessarily coming to any firm conclusions concerning what it is.* They don't even know whether it's continuous or quantized, smooth or grainy. Or as I was surprised to hear Phil Morrison say:

> When you speak about a continuous property like time or space, you are speaking about a thing that is described by an infinite number of decimal places. But nobody's ever measured an infinite number of decimal places. And nobody ever will. So it's a question whether our theory should contain this essentially unmeasurable property. It's true that it's convenient and I use it all the time, but it's an assumption that makes me a little nervous. We should not commit ourselves so thoroughly.

Sir James Jeans voiced a similar caution about conclusions springing from the theory of relativity: "It deals with measures of things, and not with things themselves, and so can never tell us anything about the nature of the things with the measures of which it is concerned. In particular, it can tell us nothing as to the nature of time and space."

We have not come close to understanding even such questions as "when" time began, for it's impossible to talk about when—a concept of time—unless you assume that you are already working in the framework of time. In a sense, asking what was before the beginning of time is like asking what's on the other side of space. Oddly, the current popular model of the origins of the universe sees the Big Bang happening *everywhere* in space, but at *one* particular time. Time has a beginning, but space doesn't.

* Although measuring something and defining it are closely connected concepts—see "The Measured Approach."

TIME AND SPACE

What is certain about time is that it can't be separated from space. Time and space are tightly woven together, not only in the extreme realms where the effects of relativity become important, but also in the familiar landscape of everyday life. A year, for example, is a *distance:* the distance that the earth moves in its orbit around the sun. If the .distance were longer or shorter (note how the space/time adjectives are interchangeable), the time would be longer or shorter, too. A day, of course, corresponds to the distance more or less around the earth's circumference—and an hour is just a fraction (1/24) of that distance. The swing of a pendulum, the vibration of a quartz crystal or atom, anything that "tells time," all inevitably also move through space. As Barnett points out, "All measurements of time are really measurements in space, and conversely measurements in space depend on measurements in time."

Space and time are so closely linked in our everyday language that we rarely stop to think about it. People say that Miami is "three hours away" from New York. If someone asks you how far it is to the grocery store, you are likely as not to answer in terms of time: ten minutes. The child on a car trip who is anxious to know how much time he has to wait before the next rest stop is likely to get an answer measured in miles. An adult knows that he or she cannot attend meetings in Detroit and Seattle on the same morning because they are separated by too much space. Of course, you could communicate by telephone at the almost instantaneous speed of light—but even then you are likely to be reminded by an annoying echo that as your conversation travels over space, it also moves through time. This is especially true if your call is carried by satellite; at those distances, even the speed of light is slow enough to make the time lag significant.*

Like the relativity of time itself, the close kinship between time and space was once considered much more natural—before it was artificially severed by the requirements of the industrial age. Noon

---

*Actually, most of the time lag is due to electronic processing of signals. Compared to our standards, light is speedy stuff.

in New York or Peking was when the sundial pointed at noon—
when the sun was highest in the sky—a measure of relationships
in space. It didn't much matter whether one town's "o'clock" happened to match another's because how would they compare times,
anyway? The trip by ship or horse or even train from one place to
another covers space and takes time, so how would you know that
the "time" at your embarkation point matched that at your destination?

All this changed, of course, with the coming of communication
at the speed of light—radio, television, modern telephones, and
so forth. Now, clocking simultaneous times at widely separated
places is not only possible but essential. In fact, the needs of television networks have been a major force behind synchronizing time:
the six o'clock news has to come on the air at exactly six o'clock all
across the country, which means that "six o'clock" has to happen at
the same time all across the country. Airline schedules, cross-
country telephone conferences, anything that forces people to
"synchronize their watches" in different places drives another
wedge in the natural affinity between space and time.

Ironically, however, it is also communication at light speed that
makes the *connections* between space and time especially dramatic. A light-year, for example, is the distance covered by light in
one year, and it is the most useful measure of distances to stars.
But it also makes it obvious that looking *out* into space also means
looking *back* into time. When you look at a star five million light-
years away, you are looking at five-million-year-old light. You are
seeing the star as it looked five million years ago. It left its source
long before human beings walked the earth. The light is only
reaching us now, but for all we know the source is long dead; the
star may be dark. The image of the star we see may be like the tail
wagging without the dog.

When an astronomer sees a quasar (or quasi-stellar object) fifteen billion light-years away at the very edge of observable space,
he or she is also seeing it as it looked fifteen billion years ago—at
the very edge of observable time. What's been happening to that
quasar during the past fifteen billion years is anybody's guess. Perhaps it cooled and a speck of it condensed into a solar system with
an earth much like ours, where humanlike beings are looking at us
the way we were fifteen billion years ago. Of course, asking what

happened "during" those fifteen billion years has no real meaning. Fifteen billion years ago on the quasar is today here. Usually, we use the term "during" to connote a passage of time, but in this case it obviously refers to a passage of space. An event that happened fifteen billion years ago on the quasar and an event that happens on earth today are strangely simultaneous.

The notion of simultaneity, however, is yet another one of the everyday absolutes that relativity did away with. Simultaneity is relative. Imagine you are sitting somewhere out in space, and along comes a large transparent room moving past you at almost the speed of light. There is a light bulb on the ceiling in the center of the room and a person sitting on a chair underneath it. Say the light flashes. The person in the room will see the light hit all four walls of the room simultaneously. But you will see something quite different: since the back wall of the room is traveling toward the light at almost the speed of light, it encounters the light sooner; you see the light hit the back wall first, the front wall later. So a determination of whether or not two events are simultaneous depends on how you are moving. And since we are all in one way or another moving continually about through space, whatever happens to us at a different time also necessarily happens to us in a different space. It's something like the strobe photographs of a moving dancer that show his motions through space as they move through time. The fourth "time" dimension is as connected to the dimensions of "depth" and "width" as the dimensions depth and width are to length.*

All this brings up the interesting question, "When is now?" Clearly, asking "when" now is makes no sense unless you also define "where" now is. The now is truly the here and now. You almost always define "now" in relation to yourself, but that may not be the same "now" for someone else in another place. Even when you look at people across the room, you are seeing them as they were a few instances ago—and if they died, like a star, in that

_____

* "The physical theory of relativity suggests, although without absolutely conclusive proof, that physical space and physical time have no separate and independent existences; they seem more likely to be abstractions or selections from something more complex, namely a blend of space and time which comprises both"—Sir James Jeans.

interval, their deaths wouldn't happen for you until their light got to you. In truth, many things don't actually "happen" for people until they "get the message"—whether the messenger brings news of birth or death, an eviction notice or a Nobel Prize.

Some cultures, says Murchie, have never tried to separate space and time. In the language of the Hopi Indians, "What happens beyond the mountain is not of us now, and cannot be discovered until later—maybe. If it doesn't happen *here*, it doesn't happen *now*. Things farther away in space must also be farther away in time."

All the various ties that bind space and time are only further pieces of evidence that seemingly isolated bits of the universe are firmly attached underneath. Space and time are linked most directly by the absolute speed of light, because light is the fastest messenger in the universe. So the three concepts fit together neatly: in order to measure speed, you need to measure distance and time—which is what speed means. But to clock speed between two distant points, you have to make sure that your clocks are synchronized. The only way to do that is to send signals via light, and still you have to account for the time it takes the light to travel. So you first have to determine the speed of light. And so on.

If it all seems curiously circular, it's only because it's all very connected. The truly curious thing is that experiment after experiment has shown that the speed of light is an absolute quantity—no matter how it's measured, or who is measuring it. No matter what the motion of the observer or the observed. No matter whether you're speeding toward the source of light, or rushing away from it.*

At the same time (or actually, a century later), countless other experiments have confirmed that measures of space and time are not absolute but depend on things like motion, or position in a gravitational field. So that the *theory* of relativity is in truth grounded in *experiment*. Indeed, the *theory* was developed in the first place in part to explain experimental *facts*. Some people think that relativity is only an esoteric set of equations of interest only to physicists and mathematicians. But even though it may not always be perceivable, relativity is a fact of life.

*See "The Measured Approach."

SPECIAL ABSOLUTISM

Actually, there are two theories of relativity: special and general. All this means is that Einstein first developed his theory to fit a special case—that of steady, unchanging motion like the motion of the ship in Galileo's example of relativity. This is "special" relativity. General relativity, on the other hand, applies to Einstein's expansion of the ideas in special relativity to all kinds of motions—in particular, changing or accelerating motions like those of objects falling under the influence of gravity.*

General relativity is all about gravity and curved space and black holes. Special relativity is all about time dilation and $E = mc^2$. Both special and general relativity are theories of absolutism because both are based on things that *do not change* in nature rather than on things that do. When you say that something is "relative," you usually mean that the way it looks depends on your point of view. The point of the two relativities is that the fundamental truths of nature look the same from *every* point of view.

The question people most commonly ask about special relativity is undoubtedly: "Can I really get younger if I travel at the speed of light?" The answer is no, you can't get younger. But you could stay young by slowing down time. On the other hand, you pay a price: you also might temporarily get more massive in the process. The relative effects that flow from the absolute speed of light and other natural laws are as follows:

First, as you travel close to light speed, time slows and space contracts. Without going into the details of the theory, you can easily imagine how something like this would have to happen: say you are traveling toward a light source (star, candle, flashlight) a million miles away at close to the speed of light—say, 180,000 miles per second. At the same time, the light from the source is traveling toward you at 186,000 miles per second. If you measure the *same* light speed whether you are speeding toward the light or standing still (and you do), then obviously something strange has

---

*Special relativity does include accelerating motions—as long as two observers are moving *steadily with respect to each other;* general relativity extends the theory to observers who are *accelerating* with respect to each other.

to happen to the space between you and the light source, or else to the time between you and the light source, or both. And it does. Even a clock sent around the world in a commercial jet liner (hardly traveling at light speed) comes home running a little bit slow compared to a twin clock left home "stationary" on the earth.

In other words, energy is equivalent to mass ($E = mc^2$*).*

Second, objects traveling at close to light speed become more massive. This is not so strange as it seems. In fact, all it means is that one kind of energy is being converted into another. Long before Einstein, people knew that the energy "potential" (or potential energy) contained in an object held up from the ground could be converted into other kinds of energy. The potential energy in a pendulum bob is converted into energy of motion as it swings, then back into potential energy at the high point of the swing, then back into motion energy, and so forth. The potential energy of an apple hanging from a tree is converted first into motion (or kinetic) energy as it falls, and then into heat energy as it hits the ground and stirs up the dirt molecules. Rubbing two sticks together to start a fire is one way of converting mechanical energy into heat energy. Einstein's quantum leap of the imagination that sprang from this notion was his realization that matter itself was a form of energy that could be converted into other forms of energy in quantitative ways. Matter, if you will, is a kind of "frozen" energy.

The conversion of matter into energy is actually an everyday

phenomenon: every time you light a fire or burn coal, you are turning the energy of matter into the energy of heat. If you were to
weigh all the molecules in the wood and in the air that make the
fire *before* you burn the wood and then *after* you burn the wood,
you would find that the ingredients get lighter in the process of
burning. But the missing matter doesn't merely go up in smoke. It

*Particles pushed to nearly light speed in giant accelerators
don't gain velocity as much as they gain another form of energy—mass. The photograph shows a small section of the PEP
(Position-Electron Project) ring at the Stanford Linear Accelerator Center.*

is transformed into a precisely measurable amount of energy, calculated by the equation $E = mc^2$, where E is energy, m is mass,
and c is the speed of light. The speed of light (c) squared is quite a
formidable number, which explains why you get so much energy at
the expense of so little mass. Nuclear bombs release a lot more
energy for the small amount of mass that they consume, because
the energy bound inside the nucleus is much more "energetic." The
sun is also fueled by nuclear reactions. It radiates trillions of tons
of its mass into space every day in the form of light energy.

The conversion of energy into mass is not so familiar, but just as frequent. In fact, every time you run, you put on a little extra "weight," or mass. And a tightly coiled spring weighs more than the same spring relaxed, due to the weight of the energy coiling puts into it. In giant subatomic accelerators, electrons pushed to 99.999 percent the speed of light gain 40,000 times their original mass—which means that the accelerators (as someone pointed out) are really misnamed: they are not so much in the business of *accelerating* the particles to high speeds as they are in the business of *building them up* to more formidable masses.

This helps to explain the always puzzling (at least for me) idea that some "particles" have no mass whatever. Massless particles tend to travel at the speed of light, and so contain all of their mass in the form of motion energy. Still, the motion energy of a light particle, or photon, is matter enough that it will fall under the influence of gravity just like any other object. A massless photon—like a massive bowling ball—would be pulled toward a flat earth in a smooth parabolic curve. The fall of a light beam is hard to detect only because the photon also travels *186,000 miles* horizontally for every second it falls.*

Finally, the energy/matter connection is the reason behind the observable fact that light speed is the speed limit of the universe. No energy or information can travel faster than light because as anything begins to approach the speed of light, it gains an ever-increasing amount of mass. Mass is a measure of inertia—a resistance to a change in motion. So the more speed or motion something has, the harder it is to make it go faster, because it also has become more massive. Eventually, the thing gets infinitely massive—which means it would take an infinite force to make it go any faster. So that even inertia is not absolute. Inertia increases the faster you go. In fact, inertia itself turns out to be a relative thing.†

---

*A falling bowling ball accelerates, or gains speed, as it falls toward the earth at the rate of thirty-two feet per second. A light particle can't accelerate because it is already traveling at the universal speed limit. Instead, light beams "falling" under the influence of gravity change frequency. A beam falling toward a heavy object like a star vibrates faster, and so looks bluer. A beam moving away from a source of gravity—moving "up"—vibrates more slowly and gets redder.
†Photons and other massless particles do travel at the speed of light, of course. But they can do so precisely because they are massless.

The reason that relativity got confused with relativism in the first place is that so many of the effects that flow from it *are* relative. That is, they look very different from different points of view. If some people are moving past you at close to the speed of light, they will seem to get very massive. However, they will not feel themselves getting massive. They will see *you* getting massive.

The energy of light—like the energy of mass—causes it to "fall" in a gravitation field. A gravitation field is equivalent to acceleration.

Stars, planets, objects, people—anything that moves quickly relative to you appears to get more massive. The increase in mass is based on motion: the greater the speed, the greater the increase in mass. But motion, as even Galileo observed, is relative. A judgment about who is moving and who is not depends on your frame of reference, your point of view. So a judgment about how massive something is must also depend on your point of view. What holds for mass must also hold for energy, of course. If the energy of motion at high speeds can be converted into matter (and vice versa), then obviously how much energy you have depends on how much motion there is. So energy in this sense is relative, too.

The same lopsided view of things also applies to time and space. Travel at high speeds causes your clocks to tick more slowly and your space to contract. But these effects are necessarily relative to something else. If your friend stays home while you whiz off on a quick trip around the stars, you may return to earth still "young" years after your friend has died of a ripe old age. But you stay young only relative to your friend and the other folks back home. According to your own internal biological clocks, or the watch on your wrist, or any other kind of timekeeping device on your starship, time flows as always. You can't tell that "time has slowed down" because even your brain is running slow, your heart and lungs are running slow. Even radioactive clocks would run slow. All kinds of clocks run relativistically because relativity is a property of time, and not of clocks. (Or perhaps I should say, *as well as* of clocks.)

The result is that you have no way of knowing whether you are more massive or not, whether your time has slowed or not, whether your space has contracted or not, even whether you are moving or not—just like a person on Galileo's ship. Your mass is normal mass, your time normal time, your space normal space. Einstein called the view which you yourself see the "proper" view, to distinguish it from other frames of reference. Everything in your view seems appropriate and proper, while things and people moving relative to you appear slow-moving, squashed, and somewhat twisted. This is certainly a familiar phenomenon on a very different front—that of comparing people's cultures: whatever "we" do is proper and normal; whatever "they" do seems peculiar and weird. I never met a parent yet who thought anyone but himself or herself knew the "proper" way of raising children. And I never heard of a national population that didn't consider its own political ideas as normal and proper, and everybody else's somewhat morally skewed. Or as Murchie reflects: "Could some such relativity of depreciation, ordinarily unnoticeable on Earth, be responsible in a slight degree for man's traditional disparagement of foreigners? Or even a hidden source of war?"

Despite all this superficial relativism, however, *what actually happens* remains remarkably absolute. Things look different from different points of view—just as a box looks different depending on whether you are standing still next to it or speeding by it in a passing car. But the box itself does not change. And the laws of nature do not change. The relationships between things and events do not change. That is why you cannot tell whether or not you are moving, whether your clocks are slow.

One of my favorite examples of this absolutism underlying relativity comes, rather unexpectedly, from the realm of biology: "Usually, we pity the pet mouse or gerbil that lived its full span of a year or two at the most," writes Stephen Jay Gould.

> How brief its life, while we endure for the better part of a century. [But] such pity is misplaced. . . . Their lifetimes are scaled to their life's pace, and all endure approximately the same amount of biological time. Small mammals tick fast, burn rapidly, and live for a short time; large mammals live long at a stately pace. . . . All mammals, regardless of their size, tend to breathe about 200 million times during their lives (their hearts . . . beat about 800 million times). . . . Measured by the internal clocks of their own hearts or the rhythm of their own breathing, all mammals live the same time.*

From the confines of our own limited perspectives, things may seem very different where in truth they are very much the same. The equations of relativity provide a kind of language—or better, a dictionary—that translates from one frame of reference into another. They allow you to begin at your point of view and move step by step to somebody else's. It is a Rosetta stone for frames of reference. Would that the simultaneous translations at the United Nations had such a facility! For what relativity really means is that everybody can agree on the facts of a situation—even though everybody sees the problem from a vastly different point of view.

---

*There are exceptions, of course, notably people—who due to a strange kind of perpetual immaturity live longer than other mammals of their size.

GENERAL ABSOLUTISM

General relativity merely extends this idea from steady motions—light beams silently passing each other in the night—to changing or accelerating motions. (Physicists use the term *accelerate* to mean any kind of change in motion—not just going faster, but also going slower, stopping, or changing direction.) The most familiar accelerating motions in the universe are those connected with the force of gravity. Falling objects fall faster the farther they fall, and planets orbiting the sun accelerate in the sense that they are constantly changing direction. The two relativities were part of Einstein's lifelong attempt to unify all of nature's forces: special relativity rested on the relative motions of electric and magnetic forces—or light. General relativity brought the final then-known force—gravty—into the family fold. The basic approach behind both relativities was much the same.

Both are "weightless."

Like special relativity, general relativity follows from what *you cannot tell*, from what *does not* make a difference. In special rela-

To an observer inside the accelerating ship, a lead ball and a wooden ball appear to fall together when released.

tivity, you cannot tell whether you are moving steadily or at rest. In general relativity, you cannot tell whether you are accelerating or standing in a gravitational field. The two situations are exactly equivalent—something Einstein understandably called the Equivalence Principle.

Imagine (as Einstein did) that you are in an elevator and the cable snaps: suddenly you are in a situation with no gravity. Drop a ball and it floats in front of you; put out your arms and they float at your sides. You are in a situation of "free fall"—no different in any way from the zero-gravity environment of outer space. As long as you are moving along with the acceleration of gravity, the force seems to disappear—just as the force of magnetism seems to disappear when you "travel along with" an electron. In this free-fall situation, is there any way you could tell whether you were in zero gravity or moving along with the acceleration due to gravity? The answer is no.

Now imagine you are floating around in a rocket ship in space, and suddenly the rocket ship begins to go faster and faster. Drop a ball, and the floor of the rocket ship will quickly catch up with it. The ball does not "float," but "falls." Can you perform any kind of experiment that will tell you whether you are really in a rocket ship accelerating in space, or just sitting in the same rocket ship in a launching pad on earth under the influence of gravity? The answer is no.

In one fell swoop (so to speak), Einstein came up with an entirely new way to think about the great eternal riddle of the relationship between inertia and mass: why do bowling balls and Ping-Pong balls fall at the same rate in a vacuum? Because if they were dropped in an accelerating rocket ship, the floor would catch up with them at the same time. Either way, the two would "hit the ground" simultaneously. If you were living in a gigantic, earth-sized accelerating rocket ship, you might think that things around you were held to the ground by the "force" of gravity. But an outside observer might see that the reason things were held to the ground was really that the ground was accelerating toward them. According to general relativity, the force of gravity is relative.

When Einstein first realized this, he was understandably jubilant. He wrote:

At that point there came to me the happiest thought of my life, in the following form: Just as is the case with the electric field produced by electromagnetic induction, the gravitational field has similarly only a relative existence. *For if one considers an observer in free fall, e.g., from the roof of a house, there exists for him during his fall no gravitational field*—at least in his immediate vicinity. For if the observer releases any objects they will remain relative to him in a state of rest, or in a state of uniform motion, independent of their particular chemical and physical nature. (In this consideration one must naturally neglect air resistance.) The observer therefore is justified to consider his state as one of "rest."

(Of course, there is a catch in this argument. Gravity pulls things toward the earth in all directions, so how could it be equivalent to accelerating in all different directions at the same time? Many aspects of general relativity are not only hard to get used to; they remain to a certain extent unsettled. Or as Weisskopf once cautioned me: "It's like the peasant who asks the engineer how the steam engine works. The engineer explains to the peasant exactly where the steam goes and how it moves through the engine and so on. And then the peasant says: 'Yes, I understand all that, but where is the horse?' That's how I feel about general relativity. I know all the details, I understand where the steam goes, but I'm still not sure I know where the horse is.")

Equivalence itself is not that hard to get used to. When astronauts accelerate toward space in their shuttles or moon rockets, they measure the force of acceleration in so many G's—G being the designation for one earth's gravity. Two G's equals twice the force of earth gravity, and so on. Visions of permanent space stations in the sky always substitute another kind of acceleration for gravity—centrifugal force. Huge circular rings swirl in space, throwing people, houses, and everything else outward like stones twirled on strings. If the "ground" is built on the inside outer edge of the rings, then the centrifugal acceleration will be exactly equivalent to gravity.

Curiously, Newton had used the example of centrifugal acceleration to prove that accelerating motion was *absolute*, not relative.

He said that while steady motion could obviously be relative (à la Galileo), accelerating motions were entirely different. If you spun a bucket of water, the centrifugal force would make the water rise at the sides, in the same way that the spin of the earth causes it to bulge outward at the equator. The effect of the bulge was clear evidence that these things were moving, and not at rest. But around 1900, Ernst Mach pointed out that if you spun the whole universe and kept the bucket or the earth at rest, then the result would be the *same*—so *still* you could not tell whether you were at rest or accelerating.

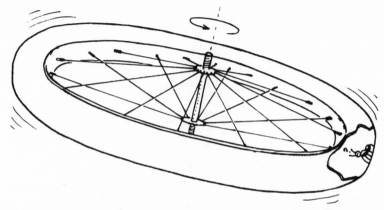

The ant inside the wheel feels a force like gravity pulling it "down"—which is outside; "up" to the ant is toward the center of the wheel.

The idea that a force such as gravity can be relative is a difficult idea to get used to. When you push something to make it go—throw a ball, for example—there doesn't seem to be anything relative about it. But consider this: one of the measures of a force is the motion it produces. Push something with a large force and it goes farther (and perhaps faster) than it goes when you push it with a small force. If motions are relative, however, and motions are the result of forces, then it is not so surprising to find that forces have relative qualities, too.

One of the major consequences of general relativity was the notion of curved space—with its ultimate expression in the exotic (yet still only theorized) entities known as black holes. A black hole is created when space bends back on itself so extremely—because of intensely concentrated gravity, if you will—that nothing, not even light, can escape. Yet all this follows from Einstein's conclusion that gravity is relative; it isn't a force that "forces" things to fall, but rather a property of space. Things fall because that is their normal "straight-line" path in a curved, four-dimensional space-time continuum. And curved space is much more than just another way to look at the force of gravity. It actually gives slightly different results which have been verified experimentally.*

The rising water in the spinning bucket was supposed to prove that some kinds of motion aren't relative after all.

Still, curved space, when you think about it, remains a funny concept. What is "straight" space? A straight line is the shortest distance between two points—but what is that? Usually, by "shortest distance," we mean line of sight, or the path of a light beam. But we *assume* that light travels in straight lines. If the light beam curves, is it going straight in curved space? Or is it curving in straight space? There are even some cases in which the shortest distance between two points is the path that takes the *longest* time according to your watch—because time *slows* as you go faster!

Relativity, in short, does not mean that everything is relative. It means that appearances are relative—and you already knew

*See "Forces, Motives, and Inertia."

that. It is not so surpising that perceptions change as you look at things from different points of view. It happens all the time. What is strange is that you can manage to reach the same conclusions from so many *different* points of view. This is something everyone has to learn. A baby looks at her toy from different points of view and sees it each time as a different object. And no wonder: a toy or a bottle *does* look very different depending on whether you view it from the top, or side, or upside down, or partly hidden under the rug. Sooner or later, however, the baby learns that the toy does not turn into something different as it presents its various faces. The more things change, the more they can stay the same.

A black hole is created when gravity bends space to extremes.

Like the baby, grown-ups often feel that reality is slippery and elusive. The scenery seems to change too often. There are too few constants that apply to all people, cultures, situations. Is there anything concrete, or "real"? Or is everything we see and know a figment of our imagination? Physics certainly hasn't helped much in this matter, turning even space and time and matter and forces into "relative" quantities. More than ever, there is nothing for people to really get a handle on. No wonder the implications of "everything is relative" are upsetting.

But relativity is only skin deep. Absolutism touches much deeper truths. Time and space may seem strangely altered in other environments—just as basic needs of people to feed, clothe, and amuse themselves take strange forms in alien cultures. Yet the things that *do not change* are real enough, and physicists have learned to see these invariants as clues to the nature of reality. Other people are learning to do the same. After all, it is hard to look at a sea of human faces and not be impressed by the similarity of expression—sadness, mirth, love, pain, pity, joy, fear, hope, puzzlement, awe, are common to all. All people value life, make music, fall in love, construct codes of law and ethics, and somehow satisfy their sense of awe at all that is the universe. Many of the same familiar features even appear in animals—our close relatives. They, too, are terrified, hungry, content. Who has not recognized something very close to home in the eyes of a dog, or horse, or dolphin?

The fact is that many things about us do not make a difference—including size, sex, and race. Most fundamental human rights and needs and feelings are—like the speed of light—invariant. In accepting them as absolutes, of course, some other things that you thought were absolute—like time and space or the righteousness of your own point of view—become necessarily relative. What is relative, in other words, depends on what is absolute. Once you have discovered what is absolute, then you can learn what is only a matter of appearance.

One day not too long ago I was sitting around with my friend the physicist talking about a book he is writing. As we pondered what possible use it could be to know all this stuff about invariants and relativity and the things that do not matter, and do not change, we somewhat laughingly stumbled upon something I remembered from Girl Scouts. Because if many of the differences among people do not make a difference, then the obvious conclusion is that we should always treat others as we would like to be treated ourselves. When I was a child, they called it the Golden Rule.

6. SCIENCE AS METAPHOR

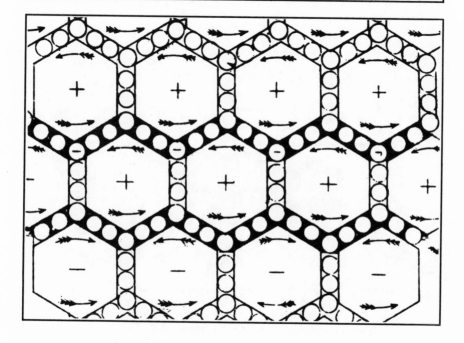

"**A**t the leading edge of experience in philosophy, science and feeling there is inevitably a groping for language to translate the insecure novelty of noticing and understanding into a precision of meaning and imagery." So wrote physicist Frank Oppenheimer in the introduction to a series of readings at The Exploratorium on "The Language of Poetry and Science." Poetry and science? Not so strange when you consider that Niels Bohr himself once wrote: "When it comes to atoms, language can be used only as in poetry. The poet, too, is not nearly so concerned with describing facts as with creating images."

Science, after all, involves looking mostly at things we can never see. Not only quarks and quasars, but also light "waves" and charged "particles"; magnetic "fields" and gravitational "forces"; quantum "jumps" and electron "orbits." In fact, none of these phenomena is literally what we say it is. Light waves do not undulate through empty space in the same way as water waves ripple over a still pond; a field is not like a hay meadow, but rather a mathematical description of the strength and direction of a force; an atom

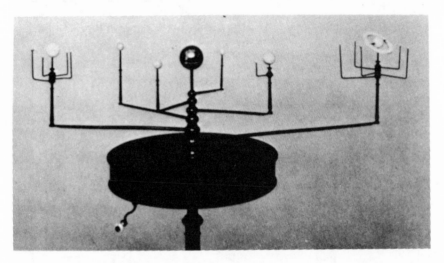

Gregory describes this orrery as "a physical model of selected features of the solar system. . . . There is a formal similarity to the planetary system which helps us to understand the solar system—provided we select what is appropriate."

does not literally leap from one quantum state to another, and electrons do not really travel around the atomic nucleus in circles any more than love produces literal heartaches. The words we use are merely metaphors; they are models fashioned from familiar ingredients and nurtured with the help of fertile imaginations. "When a physicist says 'an electron is like a particle,'" writes physics professor Douglas Giancoli, "he is making a metaphorical comparison like the poet who says 'love is like a rose.' In both images a concrete object, a rose or a particle, is used to illuminate an abstract idea, love or electron."

Over the centuries the metaphors of science have taken a multitude of various forms. Here's Nobel Prize-winning physicist Sheldon Glashow describing the world of the very small in a recent issue of *The Physical Review:*

A popular picture of particle physics has as fundamental fermions four kinds of quark color triplets and the known leptons. Strong interactions are an exact color SU(3) gauge theory; weak interactions and electromagnetism are a spontaneously broken SU(2)$\otimes$U(1) gauge theory. Three kinds of quarks (ρ,π,λ) make up observed hadrons; the fourth charmed quark ρ' is needed to explain the observed absence of strangeness-changing neutral-current pheonomena in order G and αG, and leads to the existence of charmed hadrons. The charged weak current is usually assumed to be $J\mu = \bar{\rho}\gamma\mu(1+\gamma_5)$ $(\pi\cos\theta + \lambda\sin\theta) + \bar{\rho}'\gamma\mu(1+\gamma_5)$ $(\lambda\cos\theta - \pi\sin\theta) + \nabla\gamma\mu(1+\gamma_5)$ e $+ \nabla'\gamma\mu(1+\gamma_5)\mu$.

Compare that with Francis Bacon's seventeenth-century description of heat: "Heat is a motion of expansion, not uniformly of the whole body together, but in the smaller parts of it, and at the same time checked, repelled, and beaten back, so that the body acquires a motion alternative, perpetually quivering, striving, and irritated by repercussion, whence spring the fury of fire and heat."

Or Isaac Newton's account of what we now call chemical reactions:

And now we might add something concerning a most subtle spirit which pervades and lies hid in all gross bodies, by the force and action of which spirit the particles of bodies attract one another at near distances and cohere, if contiguous . . . and there may be others which reach to so small distances as hitherto escape observation . . . and electric bodies operate to greater distances, as well repelling as attracting the neighboring corpuscles; and light is emitted, reflected, refracted, inflected, and heats bodies; and all sensation is excited and . . . propagated along the solid filaments of the nerves.

Or, finally, Christian Oersted's early nineteenth-century image
of electricity: "The electric conflict acts only on the magnetic particles of matter. All nonmagnetic bodies appear penetrable by the
electric conflict, while magnetic bodies, or rather their magnetic
particles, resist the passage of this conflict. Hence they can be
moved by the impetus of the contending powers."

The subjects of science are not only often unseeable, they are
also untouchable, unmeasurable, and sometimes even unimaginable. The only way even to begin to examine these elusive entities is
to scale them up, or shrink them down, or give them a familiar,
solid form so that we might finally get at least a temporary handle
on them. But even in 1882, physicist and lawyer John B. Stallo
recognized that the current models of the universe were only "logical fictions," useful tools for understanding but in the end only
"symbolic representations" of the real world.

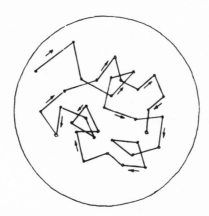

Brownian motion. A lightweight particle gets bounced around
by unseen molecules and exhibits "jittery behavior."

When it comes to science—like so many other things—we find
ourselves literally at a loss for words. Thus are metaphors born.
When Robert Brown first noticed the quick random motion of
plant spores floating in water (now known as Brownian motion),
he described it as a kind of "tarantella," according to George
Gamow,who went on to anthromorphize it as "jittery behavior."

(Brownian motion was the first convincing evidence for the existence of molecules, since it was bombardment by water molecules that made the plant spores dance.)

Later, Gamow described X rays as a mixture of many different wavelengths of invisible light. "Being suddenly stopped in their tracks [by a target], the electrons spit [sic] out their kinetic energy in the form of very short electromagnetic waves, similar to sound waves resulting from the impact of bullets against an armor plate." Thus in German they are called *Bremsstrahlung*, or "brake radiation."

Sometimes the metaphors get confused. A mixture of many colors is called white, but we also call a mixture of sounds "white" noise; we speak of "loud" colors. Something that is "going to seed" is deteriorating, yet "seedy" really means "fertile," since seeds are the origin of new growth. The universe is described alternately as a bubble, a void, or a firecracker. Time is "fluid," or "grainy," or both. Electrons are waves, and light waves are particles. If it all sounds as if nobody knows what he's talking about, it is at least in part because a lot gets lost in translation.

Imagining the unseeable is hard because imagining means having an image in your mind. And how can you have a mental image of something you have never seen? Like perception itself, the models of science are imbedded inextricably in the current world view we call culture. Imagine (if you can) what the planetary model of the atom would have looked like—its satellite electrons orbiting its sunlike nucleus—if people had still thought the earth was flat. It would have been—literally—unthinkable. "A model or picture will only be intelligible to us if it is made of ideas which are already in our minds," wrote Sir James Jeans. It was J.B.S. Haldane who noted that the inner workings of nature are "not only queerer than we suppose, but queerer than we can suppose."

Unable to suppose what the universe is really like, we rely on our rather limited but comfortably familiar models. The look of those models changes periodically, with the result that our view of the universe changes drastically. It's a long way from Newton's mechanical clockwork universe, controlled by invisible pulleys and springs, to today's image of forces as wrinkles in space, of matter as mere vibrating wisps of reality. "Scientific theories," writes Isaac Asimov, "tend to fit the intellectual fashions of our times."

Asimov goes on to detail the specific case of the atom, as good an example as any since atoms are still, essentially, unseeable—or at least require a completely different kind of seeing than the one we are used to. The Greeks, who specialized in geometry, saw atoms as differing primarily in *shape*. Fire atoms were jagged, so fire hurt. Water atoms were smooth, so water flowed. Earth atoms were cubical, so earth was solid. Along came 1800, and the world had gone metric—in the sense of being mainly interested in measuring. Shape was no longer interesting; only *amounts* mattered. Thus atoms became featureless little billiard balls, differing mainly by the quantity of mass they contained. Later still, in the 1890s, the latest fashion in science was the notion of the force field—and so atoms were seen to differ mainly according to the configuration of their outer electron clouds. All of these images persist today in one form or another, with physicists still focusing on quantities, organic chemists on the shapes of molecules, and so on.

Another familiar example of this phenomenon is plainly visible in the night sky: the stars in the Northern Hemisphere are clustered into constellations that mirror the images that danced in the heads of the Greeks who named them: all romance and adventure, the stars tell stories of queens and warriors, gods and beasts. The stars of the Southern Hemisphere, on the other hand, were named by a more modern culture whose main interest was navigation. They did not see bears and lovers in the sky, but rather triangles, clocks, and telescopes. "The division of the stars into constellations tells us very little about the stars," wrote Jeans, "but a great deal about the minds of the earliest civilizations and of the mediaeval astronomers."

Of course, it's not surprising that the way we see atoms, or stars, should change, since images of more everyday things also change drastically from time to time. Our view of childhood, of physical beauty, of women, of work, of religion, of government, all look very different in different eras. I was arguing with some friends recently that there *was* such a thing as being too thin (if not too rich) in that today's "models" were looking downright scrawny, with even *Playboy* centerfolds becoming more emaciated every year. But nothing seemed to convince them until one night we turned on the TV and saw Judy Garland in *The Wizard of Oz*. This model child of yesteryear, everyone agreed, today looked positively fat.

It is obviously not true that the magazines and movies that contain our "models" only "reflect" the real society. As in science, the relationship is a great deal more complicated. And as in science, some models (like the flat earth) are stubbornly hard to get rid of. If they weren't, it wouldn't have taken almost two thousand years to see that the earth moved around the sun—or for people to begin to abolish slavery. Even today, many cultures plant by the phases of the moon, or think of wives as property.

Metaphors, like perceptions, are drawn from common experiences. There is no way to imagine the unknown except in terms of the known, and so the landscape of the unfamiliar gets filled in mostly with familiar images. The images we use to describe both the unseeable subjects of science and the unseen future necessarily are fashioned from the "seeable" world we experience every day. And there's the rub. We do not experience the very large or the very small, the invisible forces and mathematical fields, the curvature of space or the dilation of time. We cannot crawl inside an atom, or zoom along at the speed of light. "The whole of science is nothing more than a refinement of everyday thinking," wrote Einstein. But everyday "common sense," he also pointed out, is merely that layer of prejudices that our early training has left in our minds.

Common sense is both necessary and useful. It becomes dangerous only "if it insists that what is familiar must reappear in what is unfamiliar," writes J. Robert Oppenheimer. "It is wrong only if it leads us to expect that every country that we visit is like the last country we saw." Yet this is precisely what people do. The truth is that a model—like a foreign language—isn't really useful until you can take it somewhat for granted. It's hard to speak a language fluently when you have to keep rummaging around in the back of your mind for the right word or phrase. And it's hard to understand complicated ideas when the simple ideas and assumptions that lead up to them are still tenuous and elusive. You can't learn much about atoms if you keep having to remind yourself: "Let's see. Now the nucleus is the thing in the middle. The electron is the much smaller thing that's on the outside. Is the electron the negatively charged one? Right, I remember." And so on. Being fluent means having words and ideas on the tip of your tongue. But once you become fluent in a language or in a set of ideas, you immediately internalize them to the extent that other languages and ideas sound automatically strange and foreign.

"Familiarity is soporific," writes physicist B. K. Ridley. It breeds consent to whatever models we're used to. It's a tender, powerful trap. "Consider the danger of familiarity," he goes on. "It seems clear that an object cannot be in two places at once; but an electron suffering diffraction can. It also seems clear that though size and position is infinitely variable, everything shares the same time; but, as Einstein showed, this is not so. We must check our intuitive ideas all the time."

It's not so easy to check these intuitive ideas because, well, they're intuitive! Embarking on new territory requires a fresh supply of words and images. But where are they to spring from? Often unknowingly, we keep returning to the same old well. Or as Einstein himself put it:

> We have forgotten what features in the world of experience caused us to frame [prescientific] concepts, and we have great difficulty in representing the world of experience to ourselves without the spectacles of the old-established conceptual interpretation. There is the further difficulty that our language is compelled to work with words which are inseparably connected with those primitive concepts.

In a word, language can easily turn "into a dangerous source of error and deception," as Einstein put it. Words become symbols. Of course, all words are symbols, but when we use them as a basis for model building, they can easily assume the status of graven images. And the images they represent imbue whole philosophies of thinking and seeing. All words in this sense are code words, so much so that the very use of the word "time" forces people to imagine a quantity produced by a clock, and the use of the noun "surgeon" can make it almost impossible for people to imagine the modifier "female" in front of it.* Role models are more than just

---

*See "Seeing Things."

tokens when it comes to the images engraved in our minds—just as the difference between "seedy" and "fertile" goes way beyond semantics.

Science has a special language problem, however, in that it borrows words from everyday life and uses them in contexts that exist only in realms far removed from everyday life. When I first tried to explain the newly discovered force particles* in terms of "the force you feel when you stub your toe," I found that I had stumbled upon a semantic thicket, because "force" on a macroscopic scale and "force" on a submicroscopic scale can masquerade as very different things. Physicists borrowed the idea of force from Newton's mechanics and applied it to quantum mechanics, where it was modified—at least to a layperson—almost beyond recognition. How can force have meaning in a system that barely allows for the notion of cause and effect? But still they talk about "force particles"—and we who were left back with our billiard-ball images of particles and "pushing and pulling" notions of forces stay hopelessly, irretrievably confused.

"Often the very fact that the words of science are the same as those of our common life and tongue can be more misleading than enlightening," says J. Robert Oppenheimer, "more frustrating to understanding than recognizably technical jargon. For the words of science—relativity, if you will, or atom, or mutation, or action—have a wholly altered meaning."

Many physicists are particularly uneasy these days about terms being applied to subatomic particles: "quark," for example, was borrowed from a phrase in *Finnegans Wake;* in German, it means something like cream cheese. But "quark," to most people, doesn't mean much of anything. Far worse, say the physicists, are those words that do. The subatomic world is teeming with strange species of particles bearing oddly familiar names. "Strange" is one of them. Yet particles called "strange" or "charmed" or variously "colored" or "flavored" are not in any way particularly unusual or pleasant or green or good tasting. The words are worse than nonsense (say the physicists) because they are downright deceiving.

Richard Feynman, for example, calls it "lousy terminology. One quark is no more strange than another quark. Maybe charm is

*See "Forces, Motives, and Inertia."

okay because it's so far out you know it isn't really charmed. But people think that up quarks are really turned up somehow, so it's very misleading." Victor Weisskopf concurs: "I always get the creeps when people talk about virtual particles," he says. "There is no such thing. It's a mathematical concept to describe the strength of a field." The term "virtual" refers to the very short-lived nature of such particles, but even the term "particle," Weisskopf points out, "is only there to remind you that the field has quantum effects." That's hardly what people think of when they hear a term like "virtual particle," something that sounds like a particle that's "almost" there.* Weisskopf particularly dislikes the terms some physicists use to describe the fifth and sixth quarks—"beauty" and "truth" (the first four are called strange, charmed, up, and down). "These words are really terrible because they have emotional overtones," he says. "All they mean is that we found out that there are different kinds of quarks and that they don't all behave alike."

For the same reason, many physicists don't like the term "Big Bang," for the explosive origins of the universe. It trivializes perhaps the most important moment in time, complains (for one) Robert Jastrow, making it seem "as if the Universe were a firecracker." Weisskopf sometimes calls it the Primal Bang.

One term that can make some sense is "color" as applied to the force that holds the quarks together inside nuclear particles—even though this "color" has nothing to do with our everyday notion of color. Quarks with complementary "colors" join together to form "white" or colorless particles, just as in everyday life. And the combination of "red," "blue," and "green" quarks (primary colors) can also add up to "white"—that is, a particle (such as a proton or neutron) that has zero color "charge." Unfortunately, it seems that different physicists sometimes use different color designations for the various quarks, so the metaphor begins to fall apart.

As my friend the physicist points out, however, why pick on words like "charm" and "color"? Where does a term like electric "charge" come from? Is it like a charge account? A charge in battle? (Obviously the usage "to get a charge out of" something comes from the science, and not vice versa.) We speak of positive and

---

*Of course, we do speak of virtual images of the kind that appear behind the looking glass—even though mirror images are every bit as real as any other kind.

negative electricity, when in fact there is no such thing—and if there were, the positive would be negative and vice versa. (Something with a negative charge actually has an excess of electrons, the particles of electric charge. Something with a positive charge has fewer electrons than it needs to make it neutral.) When an atom gets "excited" it does not sit on the edge of its seat (although it may dance around a bit). On the subatomic level, force means something closer to "interaction," and the strength of a force becomes the probability of its occurring. For this reason, other physicists think that worry about "truth and beauty" is kind of silly. Or as some have been heard to say, "Let the theorists have their fun."

The real trouble with words is that they automatically embody images, whether we recognize this or not. Take the word "wave," for example. It is almost impossible to think of a wave without conjuring up an image of something that looks like a water wave. And for many centuries, nobody could figure out what light was because of this linking of wave to its image in water. Water waves travel through water—more or less the way sound waves travel through air and other substances. If light were a wave, it seemed painfully obvious that it had to move through something, too. The painful part was figuring out what that something might be.

As it turned out, no one could find this mysterious substance, or even imagine its clearly impossible properties. It was called the "luminiferous ether," and from the late seventeenth century until the time of Einstein, people were as certain of its existence as other people had been certain that the earth was flat. Yet in order to vibrate fast enough to carry light, this ether would have to have the properties of a solid. Needless to say, this posed a few problems. "If the all penetrating ether is solid," writes Gamow,

> how could the planets and other celestial bodies move through it without practically any resistance? And, even if one would assume that the world ether is a very light, easily crushable solid material, like Styrofoam, the motion of celestial bodies would bore so many channels in it that it should soon lose its property of carrying light waves over long distances! This headache was pestering physicists for many generations until it was finally removed by Albert

Einstein, who threw the ether out the window of the physics classrooms.

Einstein was able to throw out the ether because he threw out the image of a light wave vibrating like a water wave. A light wave could travel through nothing at all because it is made, essentially, of a moving electric field that sets up a moving magnetic field that in turn sets up a moving electric field and so on and so forth—pulling itself up by its bootstraps. It's like an electric motor turning on a generator that turns on a motor and so on. It doesn't need to travel *through* anything because its electric and magnetic fields create each other as they zip along—at 186,000 miles per second, mind you. But it's easy to see how the image of water waves hung people up.

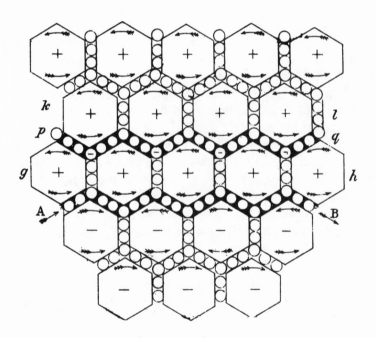

Conception of the "luminiferous ether."

There are, of course, many other examples of this throughout history. Pythagoras's model of planets revolving on invisible spheres became so strongly entrenched in Greek thought that "the Greeks soon seemed unable to imagine any planet without its orbital sphere upon which it moved in a perfect circle," writes Guy Murchie, "any other orbit being obviously less godly." More recently, Stephen Jay Gould remembers how hard it was for people to accept the idea of continental drift because it seemed so contrary to current thinking; then, once it caught on, everybody seemed to think anyone who didn't accept it was stupid.

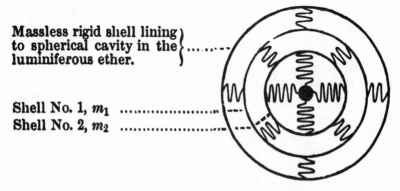

Massless rigid shell lining to spherical cavity in the luminiferous ether. ....

Shell No. 1, m_1 ....

Shell No. 2, m_2 ....

Another representation of the ether.

Even Einstein got stuck on his image of an essentially unchanging universe. Ironically, he even invented something somewhat like the notorious "world ether" to make his model work, except in this case what he invented was a mathematical device called the "cosmological constant" that would oppose the pull of gravity. Later, he called it "the greatest blunder of my life." (Aha! I can hear some people thinking. So Einstein was stupid, too. But people who believed in the existence of a luminiferous ether or a cosmological constant were no more stupid than people who can't "imagine" curved space—or women as surgeons. Our limitations are an essential part of the way we think about things.)

REDUCTIO AD ABSTRACTIUM

Models are as impossible as they are perfect—just like fashion models, superwomen, or Superman. And physics is just full of per-

fect but impossible things: ideal gases, perfect crystals, the ubiquitous bald billiard ball that serves as a model for everything from
atoms to stars. A central feature of science is "the process of abstraction," writes Philip Morrison,

> the distilling from some bit of the real world a more cleanly
> defined system that will, one hopes, still exhibit the proper-
> ties of the real in which he is interested. Much of the excite-
> ment that can be found in the practice of physical science
> has to do with seeking clever abstractions for complicated
> physical systems and then justifying the choice of the ab-
> straction.

The abstractions of science are stereotypes—as two-dimensional and as potentially misleading as everyday stereotypes. And
yet they are as necessary to the progress of understanding as fil-
tering is to the process of perception. Science would be impossible
without them—if only because the real world of nature is much too
complicated to deal with in its natural form. Abstractions are a
way to distill the essence from an otherwise unfathomable situation; some put things in focus so clearly that using them is almost
like opening a new eye on the world or, rather, like stopping down
a shutter to let the essential shine through the mass of peripheral
distractions. Raw reality is much too rich even to consider most of
the time—too various, too exceptional, too many-hued. No wonder
people so often rely on stereotypes and role models to help them
through their everyday thinking and everyday life. If you stopped
to think about all the aspects of everything (even anything) you
did, you would become completely (if temporarily) paralyzed.

"Physics is about the simple things in the universe," notes one
physicist, and yet "it could be argued that simple things plainly do
not exist." Biology and chemistry are incredibly complex compared with physics, but even such a seemingly simple thing as
a stone, he says, is "much too complicated for a physicist to
deal with."

The simpler the models, the more removed they are from reality. Yet the simplest models are often the most useful ones. That's
one reason that math is such a powerful tool in physics. It's an
ultimate abstraction that cleanly takes care of many of the messy

details of reality by temporarily dispensing with them altogether. All models, in a sense, are intermediary steps on the road to mathematical abstraction. As Richard Gregory puts it, they are a kind of "cartoon-language. Just as the pictographs of ancient languages become ideograms for expressing complex ideas—finally expressed by purely abstract symbols as pictures become inadequate—so such models become restrictive. They give way to mathematical theories which cannot be represented by pictures or models."

Today, mathematics has become very much the language of science. (Or at least, I should say, of physics.) The objects of study are mathematical and so are the models and even the metaphors. I was surprised to hear theoretical physicist David Politzer of Caltech describe the most recent "inventions" in the physics of the early universe—the moments just after the Big (or Primal) Bang —as mathematical theorems. "English is just what we use to fill in between the equations," he said. "The language we use to talk to each other doesn't have analogies in nature. But we have greatly extended our mathematical vocabulary, and we are always looking to expand this set of metaphors. That's what it's all about: Understanding is a way of picturing things, and mathematics gives you a way to do it."

This is absolutely frightening to those of us who wish physicists would become more fluent in the art of "saying it in English." Politzer, like so many others of his kind, is insisting that their discoveries are essentially nontranslatable into "our" language. But this need not be unnerving once you consider that it's impossible to experience almost anything beyond a superficial level until you learn its special language—whether it's tennis or ballet or law. Like any other "jargon," math is a vehicle that lets you go a great deal farther than you could go without it. (My friend the physicist once gave as an example of the usefulness of jargon the phrase "second cousin twice removed." Although it does not mean much unless you know the jargon of family relations, it is certainly a lot simpler to say, "Frieda is Mike's second cousin twice removed" than it is to say, "Mike is the great-great-grandson of the man who is Frieda's great-great-great-grandfather." Sailors are also well aware of the usefulness of jargon. Once I was sailing with a boatload of novices when we were about to run aground. The skipper ordered everyone to hike out over the lee rail—and about half the crew went to port while the other half scampered to starboard.)

Math is particularly useful jargon in that it allows you to describe things beautifully and accurately without even knowing what they are. You can forget about the problem of trying to imagine the unimaginable in everyday terms, because you don't need to. "The glory of mathematics is that *we do not have to say what we are talking about*," writes Feynman (emphasis his). Curiously, these mathematical images often come closer to describing reality than images fashioned from reality itself. And as many "discoveries" have been made in physics by looking at equations as by looking through microscopes and telescopes. "There is a mystery to this," says Feynman, "how mathematical thinking seems to make things fit." Unfortunately (or perhaps fortunately), we cannot make a mathematics of the world, as Feynman points out, "because sooner or later we have to find out whether the axioms are valid for the objects of nature. Thus we immediately get involved with these complicated and 'dirty' objects of nature, but with approximations ever increasing in accuracy."

It is not so surprising that when the mathematical models get dressed in the metaphors drawn from everyday experience that we get into trouble. "The history of theoretical physics," wrote Jeans, "is a record of the clothing of mathematical formulae which were right, or very nearly right, with physical interpretations which were often very badly wrong." Newton's laws of motion were almost entirely right—*entirely* right if you neglect such extreme instances as travel at the speed of light. Yet when they were interpreted as the inner workings of a giant mechanical clockwork that existed in absolute space and time, they "put science on the wrong track for two centuries." In the same way, the mathematical formulas describing the interaction of electric and magnetic fields (light) only went wrong when they were interpreted as the undulation of light waves through the world ether.

Mathematical or otherwise, our images of nature are always bound to be somewhat wrong. But even inaccurate mental models can be useful. A young physicist I know argued that it was bad to introduce people to atomic structure by letting them imagine electrons in orbit around a nucleus like planets around a sun. The model was "wrong," he argued. But all of us (including most scientists) begin the journey to the center of the atom with this comfortably familiar image; only later did physicists embellish it with the subtle complexities of quantum states. The orbits were a kind of scaffolding that helped people to get their footing while climbing toward a deeper understanding. As J. Robert Oppenheimer wrote about his "house of science": "It is not so old but that one can hear the sound of the new wings being built nearby, where men walk high in the air to erect new scaffoldings, not unconscious of how far they may fall."

Scotsman James Watt constructed a workable steam engine in the eighteenth century based on an incorrect theory of heat. A hundred years later another Scot, James Clerk Maxwell, constructed a theory of electrodynamics based on "a lot of imaginary wheels and idlers in space," writes Feynman. "But when you get rid of all the idlers and things in space, the thing is OK." P. A. M. Dirac first predicted the existence of antimatter by imagining holes in empty space. Antimatter turned out to be real enough even though the holes didn't.*

Models are stepping-stones. Just as Einstein built on the structure erected by Newton, so Newton built on that of Kepler and Copernicus. (If Copernicus had not published his treatise on the heliocentric solar system, if Kepler had not precisely calculated the elliptical paths of the planets, Newton could never have seen the similarity between their motions and the fall of the apple.) A model can serve as a solid foundation even when that model is assumed to be wrong. The very same Maxwell apparently didn't himself believe in the currently popular model of an electrically charged atom. "It is extremely improbable that when we come to understand the true nature of electrolysis we shall retain in any form the theory of molecular charges," he wrote, "for then we shall have obtained a secure basis on which to form a true theory of electric currents and so become independent of these provisional hypotheses."

---

*Although in one sense, the holes are still real and physicists sometimes speak of "holes in the Fermi sea," and so on.

SEEING THROUGH IMAGES

There is a more fundamenal reason for using models, however—even when these mental images of things are inevitably
blurry compared to the more measured terms of mathematics. We
visualize because "seeing" is inextricably linked with understanding. ("I see" is synonymous with "I understand.") Visualizing helps
us think about the unthinkable. Sometimes nature is unthinkable

*Michael Faraday with his "tubes of force." Visualizing helps us
think about the unthinkable.*

because it's so complicated. "In one chunk of ordinary material you
have 10^{23} atoms," says Caltech president Marvin Goldberger.
"Even if you had a computer that could deal with that many interactions, you still couldn't imagine it. Even if it were practical, it
wouldn't be useful. So we go back and forth, between the words
and the pictures."

At other times, nature is unthinkable because it is so far removed from our everyday experience. Trying to picture the universe before the beginning of time or beyond the boundaries of space confronts us with the unimaginable. (We may be able to think about *when* the Big Bang was, for example, but *where* was it?) To a lesser extent perhaps, the same is true of the subatomic world, and particularly of quantum mechanics.

"The magic of quantum mechanics is that we can talk about things we can't visualize," says Weisskopf. "Almost everything in science is outside your direct experience—like the earth going around the sun. But if you reduce the size of the earth and the sun and make a model of it, then you can visualize it. But you cannot visualize quantum mechanics. Our imagination is just not up to nature."

It was (and is) this inability to *imagine* quantum mechanical things that makes people so upset about them. (Merely *disagreeing* with the theory is a different thing.) During the beginnings of the quantum revolution in the early twentieth century, physicist Erwin Schrodinger said he was "discouraged, if not repelled" by the "transcendental algebra" of fellow physicist Werner Heisenberg—mainly because it "defied visualization." Heisenberg obviously understood the problem, because at one point he wrote:

> Above all, we see from these formulations how difficult it is when we try to push new ideas into an old system of concepts belonging to an earlier philosophy, or, to use an old metaphor, when we attempt to put new wine into old bottles. Such attempts are always distressing, for they mislead us into continually occupying ourselves with the inevitable cracks in the old bottles, instead of rejoicing over the new wine.

Feynman understands the importance of visualization as much as anyone—despite his repeated insistence that some things can be understood only through the language of mathematics. Feynman, after all, is a master visualizer. His Feynman diagrams are a visual language for describing complex subatomic events as a collection of simpler ones. About a year ago, I got the rare opportunity to sit down and talk with him about how he *thought* about physics. His response was, as usual, enlightening.

We learned something from Einstein. He wanted to put those two pieces together (electromagnetism and gravity, as part of the effort to find a common link among all the forces of nature), and he failed partly because he started too early. It was like trying to put together a car when you only have two pieces. Today, we have many more pieces of the puzzle, and the puzzle is much more complicated.

Feynman diagrams.

But he also failed—of course, I don't know why Einstein failed—but when he did his early work, he visualized a lot. A guy going up in a spaceship sending light back and forth (which helped him "see" the true relationship between time, distance, and the speed of light); a guy going up in an elevator (which helped him see that gravity and acceleration were equivalent). He got the idea that way, and then he formed elegant equations to explain the idea. He was good at it. It was nifty.

Later, when he was searching for a unified theory, he used a different kind of thinking—a guessing at mathematical forms. For years I've tried that and have fallen on my face. So I'm trying to do it visually. [The Feynman diagrams] aren't sufficient for this so I'm searching for a new descriptive imagery. I'm trying to follow my own advice.

The problem that Feynman was working on at the time was the nature of the mysterious force that binds quarks and their "force particles," or gluons, inside atomic particles.* He explained how he liked to take "the most odd, peculiar, striking" thing about the problem and abstract it away from the rest. In the case of quarks, the most striking thing is that the color force increases as the quarks get farther apart. "So you ask," said Feynman,

> what's the simplest way of stating it? How can you make a simpler problem with the same peculiarity? Say I use only two gluons, forget about the quarks. Assume that space is two-dimensional. If I try the theory with only two colors and no quarks and two dimensions, I think I understand why the gluons don't come apart. Now I have to climb back up and put the quarks back in and see if it still works. I may have thrown out the baby with the bathwater in trying to simplify, but I don't think I did.

Throwing out the baby with the bathwater is always a problem with building models—because models are always abstractions. One never knows for sure whether the model has really gotten to the essence of things, or just gotten rid of it. And if the kinds of stereotypes we have constructed of people are any guide, the answer is not encouraging. In most cases, we have skimmed only the surface while leaving the deeper reality untouched.

"To what extent do models help?" asks Feynman. "It is interesting that very often models do help, and most physics teachers try to teach how to use models and to get a good physical feel for how things are going to work. But it always turns out that the greatest discoveries abstract away from the model and the model never does any good."

Scaffolding is only a facade. Eventually, even the strongest scaffolding gets cast aside, and the best models are replaced by newer ones. Einstein's relativity replaced the luminiferous ether; the Bohr atom refined Ernest Rutherford's miniature planetary model by combining it with a musical image of standing waves.

---

*See "Forces, Motives, and Inertia."

Scaffolding is a great support from which to build and remodel and
fine-tune, but the trick is remembering that it's not the real thing.
The mistake most people make with role models is forgetting that
they are at best superficial caricatures of a much more complex
reality. Trying to "live up to" a model is by definition impossible—
because a model is only one aspect of a many-faceted thing.

And taking models too literally can lead us into hopeless and
unnecessary confusion. People often get frustrated in their attempts to learn about atoms because the image of everyday "particles" is so indelibly (and perhaps unconsciously) imprinted in their
brains. It's natural to want to know just where an electron is, or in
the case of radioactivity, just where the electron was hiding in the
nucleus before it was emitted. Just where are the electrons in an
atom during the transition between one quantum state and another? (Just where are they at any time?) And which electron occupies which quantum "orbit"?

But this way of thinking about electrons, Jeans points out, is
akin to believing that your bank balance actually consists of so
many coins in a particular pile. When your balance changes by a
certain number of dollars, you don't imagine those dollars actually
flying through the air from, say, your account to that of the department stores whose bills you just paid. You do not worry about
which particular dollars pay for the rent or the groceries. If you
insisted on trying to put your finger on these pieces of information,
you would hear yourself talking very much like a physicist talking
about electrons: you would have to say that which particular dollars pay the rent is largely a matter of chance.* This, says Jeans,
"may be a foolish answer—but no more foolish than the question."

Models can also be misleading when used in inappropriate contexts, say, when simple models of physics are applied to complex
things like people. As Gould points out, "a machine makes a poor
model for a living organism. Physical models often imply simple,
inert objects like billiard balls that respond automatically to the
impress of physical forces. But an organism cannot be pushed
around so easily." Yet we talk about the "force of habit" or the
"pressure to achieve" as if we were just as inert as billiard balls.
We speak of "balance of power" as if we knew how much power

*See "Cause and Effect."

weighed and how to measure it. We speak of "forcing" people or nations to do things as if we knew which buttons to push to make them go and make them stop, as if there were only a single possible response to our actions.

In the end, models and metaphors are useful only to the extent that we understand their meaning and limitations. It does not do any good to understand atoms in terms of billiard balls if we don't understand billiard balls, or motives in terms of forces if we don't understand forces. As Jeans puts it: "Nothing is gained by saying that the loom of nature works like our muscles if we cannot explain how our muscles work."

7. RIGHT AND WRONG

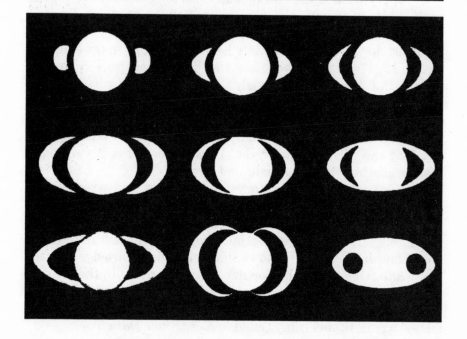

$\mathbf{S}$everal years ago, I was invited to speak to a group of "gifted" junior high school students in our community on the subject of Science and Creativity. Thinking that nothing could be quite as creative as Einstein's theory of relativity (what could be more creative than refashioning our fundamental notions of matter, space, and time?), I decided to try them out on that. All went well until the end, when a girl sitting way in the back asked: "But what if Einstein was wrong?"

What indeed? It was a fair question, to be sure. Science seems littered with the mostly forgotten remnants of "wrong" ideas, much as old love affairs are littered with false promises. Heat is not a fluid; the earth is not flat nor does it reside at the center of the universe; the planets do not revolve in perfect circles on fixed celestial spheres; Mars is not covered with canals; no luminiferous ether pervades our space, undulating invisibly as a carrier of light. On the other hand, empty space is now described as incredibly curved, and even vacuums are said to come in several exotic varieties. It seems as if the outrageous ideas of yesterday are the scientific facts of today—and vice versa. So why shouldn't Einstein be wrong?

Einstein will almost certainly be proved wrong in the long run. Or, at least wrong in the sense that he himself proved Newton wrong. But "wrong" is obviously the wrong word for it. The girl's question reminded me of a conversation I once had with M.I.T.'s Philip Morrison about whether some current views of the universe were "right" or "wrong." Finally, Morrison said to me: "When I say that the theory is not right, I don't mean that it's wrong. I mean something between right and wrong."

Unfortunately, the territory *between* right and wrong is uncomfortably unfamiliar to most of us—especially when it comes to sciences. "It's a scientific fact" is virtually synonymous with "It's absolutely true." Smearing social theories with shades of gray is one thing, but everyone knows that scientific knowledge is black and white. Or so goes the popular misconception: "In the conventional model of scientific 'progress,' we begin in superstitious ignorance and move toward final truth by successive accumulation of facts," writes Stephen Jay Gould in *Ever Since Darwin:*

> In this smug perspective, the history of science contains little more than anecdotal interest—for it can only chronicle past errors and credit the bricklayers for discerning glimpses of the final truth. It is as transparent as an old-fashioned melodrama: truth (as we perceive it today) is the only arbiter and the world of past scientists is divided into good guys who were right and bad guys who were wrong.

Anyone who has ever taken a science course knows how important knowing right from wrong is; the questions on quizzes allow only "right" or "wrong" answers, so no wonder we so often think

that the point of science is getting it "right." There are only right
and wrong answers to such engineering questions as How much
weight can be carried by such-and-such a bridge? But it turns out
that very little in science is actually wrong—and nothing in science
is ever completely right.

Take Isaac Newton, for example. There is no argument about
the fact that Einstein proved Newton wrong. Newton said that
time and space were absolute, and Einstein proved they were not.
Newton never conceived of gravity as an unseen curvature of
space. Newton didn't realize that mass was a form of energy, or
that inertia would become infinite as you approached the speed
of light.

Yet Newton's "wrong" ideas still chart the paths of space shuttles and place artificial satellites into nearly perfect orbits. Apples
still fall and the moon still orbits according to Newton's formulas.
For that matter, Newton's theories work well for everything in
our daily experience. They break down only at extreme velocities
(approaching the speed of light), where relativity comes into play,
or at extremely small dimensions, where quantum theory takes
over, or in the presence of extremely massive objects such as black
holes. "Einstein's correction of Newton's formula of gravity is so
small," writes Arthur Koestler, "that for the time being it only
concerns the specialist." Einstein's equations even give the same
answers as Newton's equations for the things that Newton was
dealing with.

Einstein proved Newton wrong only in the sense that he stood
on Newton's shoulders and saw things that Newton could not
see—like what happens to time and space under extraordinary (to
us) conditions. Mostly, Einstein proved Newton *right*, since his
theories were built on Newton's foundations. Einstein took Newton's ideas and stretched them to previously unimagined limits,
brought them into a new dimension, made them broader, bolder,
more sophisticated. Einstein added to Newton just as today's
physicists are adding to Einstein. Einstein climbed the tower of
Newton's scaffolding and saw things from a better perspective. If
the scaffolding hadn't been strong, he would have fallen flat on his
face.

Right and wrong turn out to be surprisingly *unscientific* ways
of describing ideas—especially scientific ideas. Rarely do revolu-

tionary concepts overthrow old ways of thinking in unexpected coups. Physicist Hendrik Casimir goes so far as to argue that no sound theory is *ever* completely refuted: "There is no 'stage of refutation,' but there is all along a process of demarcation and limitation," he writes in his book *Haphazard Reality*. "A theory, once it has reached the technical stage, is not refuted, but the limits of its domain of validity are established. Outside these limits new theories have to be created."

Or as British physicist David Bohm put it: "The notion of absolute truth is shown to be in poor correspondence with the actual development of science. . . . Scientific truths are better regarded as relationships holding in some limited domain."

New ideas expand, generalize, refine, hone, and modify old ideas—but rarely do they throw them out the window. Some "wrong" ideas are misconceived, or wrong only in that they are awkwardly formulated. Some turn out to be not so much wrong as unnecessary or irrelevant. Like the luminiferous ether, or James Clerk Maxwell's "wheels and idlers in space," or the notion that heat is a fluid, new theories render these constructs superfluous. But the misconceptions at the root of most "wrong" ideas in the history of science are sins of omission: they were wrong because they failed to take something into account, to see some part of nature that was keeping itself invisible, to notice connections among things that seem on the surface totally unconnected. "Wrong" more nearly means "limited."

For centuries, people argued over whether the wave theory of light or the particle theory of light was correct. But light turned out to be both: part wave and part particle. Both theories were right, but restricted. A correct theory requires aspects of both.*

Even the idea that the earth is flat was largely the result of a limited outlook at our large, spherical planet. The earth certainly *seems* flat enough as you walk around town. But the view from home is always somewhat parochial, and the earth doesn't begin to look round until you get far enough away from it. Today, most people have seen the spherical earth in its true shape and colors in images brought back from orbiting satellites. Yet hundreds and even thousands of years ago people like Columbus and Erat-

*See "Natural Complements."

osthenes were able to see much the same view with the aid only of their imaginations. Physically or intellectually, the difference between a round earth and a flat one is primarily one of perspective—a broad versus a narrow point of view. Space-time itself only begins to look curved when your measurements cover a large enough territory. And quantum mechanics and relativity are merely ways of offering larger perspectives on classical ways of viewing things.

As Einstein described it, constructing a new theory is not like tearing down old buildings to erect new skyscrapers. It is rather like climbing a hill from which you can get a better view. If you look back, you can still see your old theory—the place you started from. It has not disappeared, but it seems small and no longer as important as it used to be.

Yet how easy it is to gloat over the wrong ideas of other people! Many years ago, I was asked to write an article about an amateur inventor who thought he had invented a "reactionless" space drive—that is, a rocket that pushed off into space without pushing off anything. This would certainly be a marvelous invention, for it would mean that rockets wouldn't have to carry the huge amounts of heavy fuel they need to "push back" on space so that the reaction can propel them forward. It would also be in clear violation of Newton's Third Law: that every action produces an equal and opposite reaction. The idea was silly enough; but what really struck me was the way the inventor was reveling in the belief that he had proved Newton wrong! (This was no big deal, of course, since Einstein had already proved Newton wrong—but the pleasure this fellow was getting out of it seemed totally out of proportion.)

It reminded me of the time my son came home from school and announced that his friend was stupid because he believed that the earth was flat. Even adults tend to equate "flat earth" with (at the very least) backwardness. And it does seem silly to think of the earth as a large pancake of a platform floating in the center of space. But isn't it even sillier to suppose that we live on a great spinning ball, and that people in China are hanging upside down by their feet in thin air?

Laughing over other people's wrongs comes largely from taking what we know for granted. Once, in a science class for first-graders, I heard a teacher casually mention that feathers and rocks would fall at the same rate in a vacuum—as if it were the most obvious thing in the world. Not a hint that it had taken humanity thousands of years to even notice it (much less to explain it).

The following week, another teacher explained to the children how sand and soil came from crushed rocks, and life came from the soil. I later mentioned to the teacher that the children might misunderstand the notion that "life comes from soil." And she said: "Oh, they know what I meant." But the truth is that if you go and look at the soil or the sand with a magnifying glass and see that it's teeming with life, you naturally come to the conclusion that life springs from soil. To say that every bit of life has to come from another bit of life—not to mention that the instructions for this life are encoded in submicroscopic spiral strands of DNA—sounds utterly fantastic.

Today, we know all about falling rocks and round planets and DNA. The previous ideas that people held were clearly wrong— just as it was wrong to think of heat as a fluid instead of a form of energy, or to imagine that planets and stars needed a constant push to keep them going. But this is hardly something to gloat about.

People do not call Ernest Rutherford a dummy, even though after he discovered the atomic nucleus* he went on to insist that anyone who saw something practical in its application was "talking moonshine." If you call Rutherford wrong, then you have to say that anyone who cannot see clearly into the future is wrong, because "right" becomes synonymous with "clairvoyant." So to gloat (or even worry) over the finding that Newton (or Einstein) might be wrong seems somewhat silly. *Of course* they were wrong. Neither Einstein nor Newton could resolve every unanswered riddle, or foresee every possible consequence of every conclusion. They did not (could not) claim to be all-seeing or all-knowing. People who do claim to possess this kind of knowledge are not in the business of science—because right and wrong in that sense are not questions of science. They are only matters of dogma.

In fact, science never proves anything completely right because there is so much left to be learned. "Each piece, or part, of

_____
*See "Seeing Things."

the whole of nature is always merely an *approximation* to the complete truth," writes Richard Feynman, with his own italics. "In fact, everything we know is only some kind of approximation, because *we know that we do not know all the laws yet.* Therefore, things must be learned only to be unlearned again, or, more likely corrected."

Or as Sir James Jeans put it: "In real science, a hypothesis can never be proved true. If it is negatived by future observations we shall know it is wrong, but if future observations confirm it we shall never be able to say it is right, since it will always be at the mercy of still further observations." (Students who take physics courses often are asked to prove things—like "Prove Ohm's Law"—as if one could really prove such a thing in the course of a three-hour experiment!)

To be sure, scientists are people, and as such enjoy an aura of "rightness" as much as anyone. But from Aristotle to Einstein, the tenets of the greatest thinkers were often held much more tentatively than popular histories have acknowledged. Newton, for example, never regarded his theory of gravity as "right." As Einstein remarked on the two-hundredth anniversary of Newton's death: "I must emphasize that Newton himself was better aware of the weaknesses inherent in his intellectual edifice than the generations of learned scientists which followed him. This fact has always aroused my deep admiration."

Politicians and journalists and social scientists are not so apt to "admire" others for admitting their mistakes; on the contrary, the admission that even part of a policy or theory is wrong is frequently touted as proof that it was (and is) completely without merit. When it comes to metaphysics at least, fixing the blame for wrong and the credit for right become almost obsessions. People say that everything from welfare to national defense, from sex education to pornography, is clearly either right (or at least okay), or wrong (or not okay) because some aspect has been misconstrued or gone awry or been vaguely associated with something useful.

In fact, the rightness or wrongness of scientific ideas tends to become tinged with dogma precisely when those ideas enter the realm of philosophy. And no wonder: categorizing ideas as cleanly right or wrong may not be scientifically useful, but *philosophically* it is immensely appealing. No one likes being left in an intellectual purgatory. And so the slow evolution of scientific theories is rewritten as a series of revolutionary coups:

"Scientific revolutions are not *made* by scientists," writes Casimir.

> They are *declared* post factum, often by philosophers and historians of science. . . . The gradual evolution of new theories will be regarded as revolutions by those who, believing in the unrestricted validity of a physical theory, make it the backbone of a whole philosophy. . . . Physics may even feel flattered by this homage, but it should not be held responsible for the unavoidable disappointments.

Even in science, of course, some ideas are righter than others. But how do you tell which is which? Right ideas seem to be those that lead to further investigation, to whole new categories of questions, to an even more passionate quest for knowledge. Right tends to open our eyes, wrong tends to close them. In this sense, Newton was right, but someone like Aristotle was wrong, because (as George Gamow puts it): "His ideas concerning the motion of terrestrial objects and celestial bodies did probably more harm than service to the progress of science." Galileo, among others, spent a lifetime trying to right Aristotle's wrong ideas about the immutable heavens, the geocentric universe, and so on.

But even this interpretation can be open to question. One of the things Aristotle was most "wrong" about was his assertion that all bodies naturally stop moving if they are no longer being pushed by a force. But in a passage I was surprised to find in a physics textbook, author Douglas Giancoli states:

> The difference between Galileo's and Aristotle's views of motion is not really one of right and wrong. . . . Aristotle might have argued that because friction is always present, at least to some degree, it is a natural part of the environment. It is therefore natural that bodies should come to rest when they are no longer being pushed. . . . Perhaps the real difference between Aristotle and Galileo lies in the fact that Aristotle's view was almost a final statement; one could go

no further. But the view established by Galileo could be extended to explain many more phenomena.

Right ideas are seeds that flower into righter ideas, whereas wrong ideas are often sterile and do not bear fruit. Right ideas have deep roots that often reveal surprising connections among seemingly unconnected things, and have an uncanny knack for sprouting the unexpected. Once Newton got the right idea about gravity, he explained a great deal more than falling apples, or even the orbit of the moon. He tied together the universe with one cosmic force in a way that allowed later astronomers to understand the motions and masses of all the stars and planets.

It is in this sense that right ideas allow scientists to "predict" novel phenomena: this kind of prediction doesn't imply looking into the future, but it does tell people where to look in the present to find, for example, radio waves. Once Maxwell got the right idea that visible light was an electromagnetic vibration, then it was not so farfetched to think that there might be other identical but "invisible" vibrations with higher or lower frequencies. The lower frequencies are microwaves and radio waves. The higher frequencies were later recognized as X rays.

Many scientists say that these *connections* are as good a guide to rightness as anything—especially when it comes to drawing lines between science and the so-called pseudosciences such as astrology. The idea that the positions of the planets may influence your day-to-day life simply doesn't fit in with anything else people know about gravity or other aspects of nature. Any idea that seems completely unconnected with the rest of knowledge is usually greeted with suspicion.

This interpretation of "right," of course, is a lot more loose-ended than the kinds of right answers most people are used to—especially from science. Right turns out to be a risky business. As Morrison says of his own work:

> The whole business contains a certain amount of danger and a high tolerance for ambiguity. You get one piece of the puzzle and you try to make it fit, or re-form the puzzle around it. But this is always with the assumption that the whole construct may be wrong, that there is no right an-

swer, that these are all ways of looking at things which are useful to different extents, and that allow you to go to the next step.

Right ideas are stepping-stones. They do not require—or even imply—100 percent accuracy. Therefore Newton (or Darwin or Freud) can still be considered essentially "right" even though his theories have undergone substantial modification. Freud was probably wrong in that he overemphasized the importance of early sexual experiences, but his far greater contribution was his essentially right and quite revolutionary recognition that *any kind of early experience* could influence adult emotions and attitudes. The creationists who argue that Darwin's supporters can't be "right" because, after all, they even argue among themselves are missing the point. Updating a theory in light of new knowledge doesn't destroy its credibility—quite the contrary. The better an idea holds up to the rigors of conflict and change, the more likely it is to prove "right" in the end. (There's probably no better example of this than the idea of democracy, a political system that incorporates mechanisms for airing conflicts and producing changes as part of its very structure.)

Realizing that there's a *better* way to be "right," in other words, does not necessarily mean that everything previous is wrong. And this is an area where science has produced an especially sweet "sentimental fruit," as my friend the physicist would say. Evolution is often a sounder foundation for progress than revolution. Just because something needs to be modified or retooled doesn't mean that all its premises are mistaken. Yet this is precisely what we usually assume when it comes to social ideas. If an overbloated bureaucracy badly administers an inoculation program for young children, wasting money in the process, the answer is to throw the program out—throwing out the babies with the bureaucracy. If a feminist leader like Betty Friedan revises her ideas in light of changing social and economic realities, people cite it as "proof" that feminism was a bad idea in the first place (and some feminists cry "Foul play!"). Newspapers are full of pronouncements about who was right and who was wrong—as if predicting the future was the point of things. Far more useful is the resiliency to adjust when ideas turn out to be somewhat wrong or in need of modification—as they almost always do. If social and political science has anything to learn from physical science, it is undoubtedly that the way to build new ideas and institutions is *on top of old knowledge*—not in its ashes.

"When we find out something new about the natural world this does not supersede what we knew before," said J. Robert Oppenheimer.

> It transcends it, and the transcendence takes place because we are in a new domain of experience, often made accessible only by the full use of prior knowledge. . . . Thus what has been learned and invented in science becomes an addition to the scientist, a new mode of perception. . . . A perpetual doubting and a perpetual questioning of the truth of what we have learned is not the temper of science. . . . The old knowledge, as the very means for coming upon the new, must in its old realm be left intact; only when we have left that realm can it be transcended.

A society or a science can accumulate a righter and righter (or more and more useful) perspective on things by building on a foundation of partially wrong half-truths. So can an individual. And when it comes to learning science, the process is almost inevitably one of replacing one "wrong" idea with another. But these ideas aren't any wronger in their context than Newton's ideas of gravity were wrong in his context. In fact, when it comes to explaining the essence of scientific ideas in everyday language, Victor Weisskopf likes to say that "you always have to lie a little to tell the truth." Some science writers get their feathers ruffled at this thought, but in my experience it's true.

For example, Weisskopf often talks about how amazing it is that the difference of a single electron can make such a big difference in the nature of an element. Neon, with ten electrons, is a chemically inactive gas. Sodium, with eleven electrons, is one of the most chemically active metals. Why is this so? The reason is basically that neon's ten electrons form a completely closed shell around its nucleus and leave the atom so self-contained that it has no inclination whatsoever to interact with anything else (which is why it and others sharing its place on the periodic table of elements are called "noble gases"). Sodium, on the other hand, has one extra electron buzzing around just itching to interact with practically anything that comes along. This extra electron is common to metals—and it is these free electrons that so freely conduct electricity.

I know a very bright young science writer, however, who insists that Weisskopf is "wrong" and his example should not be used. Weisskopf is wrong, says the science writer, because the only difference between sodium and neon is *not* one electron; sodium also has an extra proton in the nucleus. And of course, he is technically right. But he is also missing the point. Nuclear particles have almost nothing to do with chemical reactions, which are determined solely by the configuration of an atom's outer electrons. If we followed the young writer's vision of truth, we would never be able to say much about nature at all, because *everything* one says about science (about anything, for that matter) is always partially false—especially if you judge something false for the crime of being incomplete.

People often shy away from science precisely because they are afraid of being wrong. Somewhere along the line they have been led to believe that all scientific questions have clear, unambiguous answers. They have been taught that science is all work and no play, all knowledge and no wonder, all logic and no guesswork. Unfortunately, this attitude excludes them not only from the point but also from the fun of the game.

In the end, the importance of being wrong is greater than one might think—because a well-thought-out wrong idea serves as a basis of comparison, a springing-off point, for right ideas. Even the ubiquitous bald billiard ball is invaluable as a model precisely because of its obvious wrongness: "We know from the outset that it is wrong in the strict sense that it cannot possibly be true," says B. K. Ridley, "and so an assessment of how wrong it is in the particular case can begin straightaway."

Weisskopf tells a story about the impatient German tourist who asks why the Austrians even bother to publish railroad schedules, when the Austrian trains are never on time. The Austrian conductor answers: "If we didn't have timetables, we wouldn't be able to tell how late we are." Many scientific models, says Weisskopf, are like Austrian timetables. (Weisskopf can tell this story because he was born in Vienna, and is still an Austrian citizen.) Some of them are partly wrong and some of them are very wrong. "But what's interesting is to see *how* and *why* they are wrong," he says. "You always need the timetables."

8. THE MEASURED APPROACH

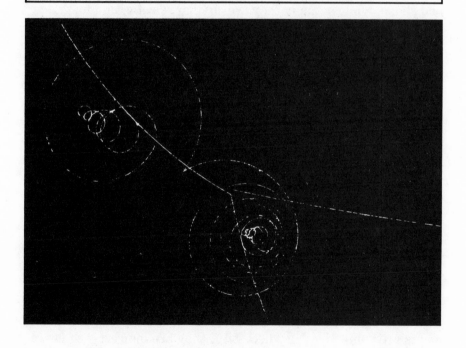

"The only way a scientist can start to understand something is to describe it, to measure it and name it." So wrote paleoanthropologist Donald Johanson in his best-selling book *Lucy: The Beginnings of Humankind*. But he might as well have been writing about teachers or economists or artists—or even judges in beauty contests. For the way we think about something determines how we size it up, and also *depends on* how we size it up. You can't begin to measure something until you make some assumptions about what that something is. You can't measure the amount of beauty in a contestant without defining in some sense what beauty is—any more than you can measure success or distance or motion or time or intelligence without defining what they are. Measurement is as much a mode of perception as seeing and hearing. It is a sharp tool and also a dangerous one. Reality and the way we measure it are tied in so many knots that it's hard to untangle the axioms from the inferences, the assumptions from the conclusions. We imagine our yardsticks to be straight, rigid, and sharp, while actually they are circular, elastic, and fuzzy.

All measurement begins (and in the end, ends) with ourselves. This may have been more obvious in the days when a foot was literally the average length of the feet of the first twelve men to emerge from church on Sunday. (To measure a room as ten feet long didn't tell you as much objective information about the room as it told you that ten human feet would fit in it.) But even today's high-tech measuring tools mostly measure aspects of themselves: clocks tell their own time, rulers measure their own inches, thermometers take their own temperatures. As Werner Heisenberg himself said, "We have to remember that what we observe is not nature herself, but nature exposed to our method of questioning."

This may seem, on the surface, outrageous. After all, what is more scientific than measurement? People who work at particle accelerators, for example, routinely make precise measurements of such esoteric things as the "emptiness" of a vacuum, the mass of minuscule subatomic particles, the energy of electron beams traveling at 99.999 percent the speed of light, the lifetimes of entities that "live" for mere trillionths of a second. Surely such measurements are "objective." By necessity, they are indirect—as far removed from the human senses as we can imagine. Lifetimes of particles are measured by the lengths of the tracks they leave in bubble chambers. The energy of a beam is measured by the amount it deflects when placed in a magnetic field. Vacuums are measured by such indirect indicators as how fast a hot filament cools off in them (the fewer air molecules present, the longer it takes to carry the heat away so the longer it takes to cool) or electrical conductivity (the current moves better the more matter there is in the vacuum).

Yet the meaning of all these measurements remains somewhat obscure, if only because they involve such murky concepts as time, space, and temperature. How do you measure time, for example? "Like other physical quantities," wrote physicist Hermann Bondi, "time is defined by the means used to measure it." A pendulum clock measures the period of the pendulum swing; an electric clock measures the frequency of an alternating current. An atomic clock might measure, for example, the frequency of vibration of the hydrogen atom. What the clock actually measures is the relationship between events taking place within the atom, and nothing else. It is as much a definition of what we mean by time as it is a measure of it. In the same way, a clock based on our planetary year measures only the relative positions of the sun and the earth. We can hardly define seconds and hours in terms of this year, and then use the same seconds and hours to measure the same year. If we do, all we learn is whether or not our clocks are accurate—certainly not "how much time" is passing.

The way people think about time has changed substantially as the means to measure it have changed. In the age of sundials and water clocks and sand-filled "hour" glasses, time "flowed." It was a continuous, fluid property. The invention of precision mechanical clocks, however, instilled the notion that time is a succession of extremely short intervals. Modern time does not flow so much as tick. (It would not be surprising if the increasingly widespread use of digital clocks produced a similar revolution: time will no longer mark the passing of the hours around the face of a clock—like the earth passing through her daily orbit—but rather it will blink at us as a sequence of numbers, a kind of continual countdown.)

The same is true of space. Textiles were called yard goods because they were goods that were measured by the yard. Even today's "scientific" meter is defined in terms of something else: the meter is precisely 1,650,763.73 wavelengths in a vacuum of the radiation corresponding to the transitions between the levels 2_{p10} and 5_{d5} of the krypton-86 atom.

Obviously, taking precise measurements requires a thorough knowledge of nature—which is one reason that the attempt to measure things has led to so much invention and discovery. Another reason is that they allow us to *compare* things. It is said that Galileo compared the time it took for a chandelier to swing back

and forth against the rhythm of his own pulse.* A sundial can compare the difference between the hours of daylight in various seasons. Measurements allow us to compare such unrelated things as
atomic vibrations and planetary orbits, the rhythm of heartbeats
and the gestation periods of babies. Intelligence tests even allow
for comparisons of the amount of intelligence various people possess: but here the standards of measurement are far less accurate
or meaningful than Galileo's measurement of the swinging chandelier. For at least Galileo knew that his standard was his own
heartbeat. We measure intelligence, however, without having the
slightest notion of what intelligence really is. Normally, we define
intelligence as that quality (or quantity) which produces good results on intelligence tests. This allows us to conclude a good deal
about the tests, perhaps, but virtually nothing about "how much
intelligence" is possessed by those taking them—for we have already defined intelligence in terms of success on the test.

The ability to measure things accurately has taught us many
things, but none more important than the limits on measurement
itself. All measurements are relative, for example. If you measure
your driving speed relative to the other cars on the road, you
might come to the conclusion that you are traveling at a reasonable
pace, while a look at your speedometer might tell you otherwise.
Both measurements are equally correct—and of equally limited
value. Each, in fact, is a measurement of a very different thing—
both called "speed" and yet obviously in need of further clarification. (I once got a speeding ticket because a traffic cop was
measuring my speed relative to the road while I was measuring it
relative to the other cars.)

Measurements also are limited by the available instrumentation. If you need to go over something with a fine-tooth comb and
you have only a coarse-tooth comb, then you are sure to miss much
that you are looking for. Often, our instruments are either too fine
or too blunt for nature. In order even to detect something, an instrument has to be honed to exactly the right scale, just as a radio

---

*Galileo's still rather surprising discovery was that the period of the pendulum—
that is, how long it takes to swing from one side to another and back—does not
depend on the size of the swing. It depends only on the length of the pendulum.
See "Waves and Splashes."

has to be tuned to precisely the right frequency to pick up a particular signal. We cannot see (much less measure) atoms with ordinary light for the same reason that an ocean wave does not "see" a piling. To "see" or to detect something requires that you are affected by it in some way—and an ocean wave is simply too large to be affected by a mere piling. The same ocean wave, however, would easily "see" something as large as a reef, because the reef is massive enough to make the ocean wave bend (that is, to affect it). Atoms are invisible to us because the light we see by is as large compared to an atom as an ocean wave is to a piling.

Crude measuring devices also make it impossible to observe aspects of human behavior. Intelligence tests are an obvious example, but so are even such seemingly straightforward measures as the TV Nielsen ratings. As a measuring device, the Nielsens particularly irk someone like experimental physicist (and Nobelist) Edward Purcell because, as he told me, the sample used is so small—so crude—that "they are making decisions about programs based on what is within the statistical [room for] error."*

Yet Purcell brings up the Nielsens mainly as an example of a deeper problem in measurement: the effect of measurement on the subject itself. "The qualitative problem is that the sample knows it's a sample," he says. The Nielsen families *know* that they are Nielsen families, and consciously or unconsciously skew their behavior accordingly. For this reason, trials of new drugs are always (or should be) "blind" trials—because it is well known that the mere knowledge that you are taking a wonder drug may be enough to make you well again. In the same way, test-taking always affects the test-takers—often to the extent that tests tell more about reactions to testing than anything else. Or as J. Robert Oppenheimer put it, efforts to probe human phenomena are "irretrievably altered by the very effort to probe them—as a man's thoughts are altered by the fact that he has formulated and spoken them."

Scientists often try to isolate their subjects from influences that might affect measurements. In this sense, scientific observation is very different from everyday observation. We normally see things in context, but a scientist quite often wants to see things *out* of

---

*The Nielsen Company, of course, disputes this.

context—in sterile environments cleansed of messy interference from the outside world. And yet the very fact of isolation leads to inevitable complications. Living things isolated from their environments die, or, at the very least, become radically altered. This posses particular problems for social scientists—for if they sufficiently isolate their subjects to eliminate extraneous outside influences, they also change their behavior to the extent that the test no longer tells the scientists anything useful about normal people. As one psychologist complained, research into the separate spheres of influence of the right and left sides of the human brain has been hampered by a scarcity of subjects with split brains. "And even these studies are of limited value for learning anything about normal people—because that is not the way most people walk around."

(There's probably no such thing as truly "objective" journalism for the same reason—unless you include *Candid Camera*. Nobody talks normally to a reporter, much less in front of a camera: people inevitably change their behavior in the same way that the subjects of scientific study change in the light of observation. Journalists are also inevitably involved in the creation of the events they report, just as physicists are involved in the creation of the particles they measure. This doesn't make the discoveries of either journalists or physicists invalid. It merely means that the significance of the discoveries must be carefully interpreted.)

It may be self-evident that people make elusive subjects for measurement. But what about things? Does a chair or a star or an atom alter its form merely because it's measured? It turns out that it does. Take temperature, for example. The minute you stick a thermometer in someone's mouth, you change that person's temperature. If you measure the temperature of a roast or a cup of soup, you change its temperature, too. That is, the way you "take the temperature" of something is by reading the indicator on the thermometer. Therefore, you are really taking the *thermometer's* temperature. This will also tell you the temperature of the thing you want to measure once the temperature of the thermometer and of the object are the same. But in order for them to reach the same temperature, heat must be exchanged. If the thermometer is colder than, say, the roast, it will draw heat from the roast, making the roast colder. If the thermometer starts off hotter than the roast, it will give some of its energy to the roast. As Isaac Asimov points out, "This difficulty can be circumvented if the thermometer happened to be exactly at the temperature of the [roast] to begin with. But in that case, how would you know the right temperature to begin with unless you measured it first?"

The real problem is that "we have to use one bit of the universe to measure another," as B. K. Ridley puts it. Measurements always require that the measurer and the measuree somehow interact. When things interact, they exchange energy, so that the act of measuring creates a new situation that is different from the one you set out to measure in the first place. On one level at least this means that objective and accurate measurements are even theoretically impossible. "Every intervention to make a measurement, to study what is going on in the atomic world," writes J. Robert Oppenheimer, "creates, despite all the universal order of this world, a new, a unique, not fully predictable situation."

Oppenheimer, of course, was writing about the world of atoms. These unavoidable errors in measurement do not trouble us when we try to measure, say, the length of a tabletop—no matter how many molecules we might inadvertently rub off in the process. With the proper instruments, you could measure a milligram of dust to within a trillionth of a centimeter. But there are two realms in which the interaction between measurer and measuree (observer and observed) take on overwhelming importance: the realm of people and the realm of atoms.

In the realm of atoms, these innate difficulties of measurement are most often described in terms of the Heisenberg relations—or the Heisenberg "uncertainty principle." Heisenberg himself visualized it this way: If you want to measure the exact location of an electron and also its exact motion, you run into a problem. In order to measure it, you have to "see" it somehow. Say you shine a light on it. If you use low-energy light, you will not disturb the electron too much, but the wavelength of the light is so long that it will not be able to define the electron's position. (Like using a coarse-tooth comb when you need a fine-tooth comb, or an ocean wave to "measure" a piling.) On the other hand, if you use high-energy light (a fine-tooth comb), you can determine the position of the electron, but the light will give it such a jolt that you will change its motion.

Many people have taken the Heisenberg principle to mean that nature is innately unpredictable and unknowable.* But nature may be innately unpredictable even without Heisenberg (or even without limiting ourselves to things as small as atoms). In order to predict what something will do, you have to be able to measure its initial position precisely. So measurement and predictability go hand in hand. Before quantum mechanics and Heisenberg, people felt sure that if they wanted to, they could measure the position and motion of every particle in the universe and therefore "predict" the future. But this greatly overstates the abilities of classical physics. For one thing, you can't resolve a piece of information smaller than a single photon of light, so the picture is always blurred in some respects. And our instruments are always affected by the inevitable Brownian motion of their molecules—the random motions that add up to the quality we call heat.

In addition, there is the problem of calculation. Philip Morrison (borrowing from Feynman) uses the example of a waterfall: classical physics says that you should be able to predict exactly where each particle of spray will go just by watching it fall over the lip. "But who's done that calculation?" he asks. "The fact of the matter is, you couldn't do it—because it's easy to show that in order to make it work you'd have to put so much information into your computer that just the gravitational effects of your computer being positioned one way or the other would make a difference in the result."

The most obvious example of the limits of measurement applied to people is, of course, intelligence tests. Almost nobody mentions intelligence tests without citing the famous example of the tests performed by the U.S. Public Health Service in 1912 on immigrants at Ellis Island. According to the tests, 83 percent of the newly arrived Jews were feebleminded—along with 80 percent of the Hungarians, 67 percent of the Russians, and 79 percent of the Italians. These results may seem ridiculous today. They obviously measured language and cultural differences that had little to do with intelligence. The art of intelligence testing has certainly been refined since then. But still it is impossible to probe another person's mind without inserting your own internal prejudices. Cultural biases are the bane of intelligence tests—for it is impossible

---

*See "Quantum Leaps."

to define intelligence without referring to some cultural standard. To say this changes the outcome is to say the least. Stephen Jay Gould recently devoted a whole book (*The Mismeasure of Man*) to show how measurements (in this case of the human skull) are molded by the (sometimes unconscious) actions of the measurers. For many decades, these dubious measurements of human skulls were used to support racist viewpoints. "Today," writes Gould, "these contents stand totally discredited. What craniometry was to the nineteenth century, intelligence testing has been to the twentieth."

Many things in life will always elude precise measurements—no matter how hard we try to sort them out or pin them down. Yet ironically, the things people have *not* been able to measure have been at least as instructive in unraveling the nature of things as the things they have measured. The Heisenberg uncertainty principle, for example, established a universal unit of smallness in the universe. Things cannot be infinitely divided, and so energy and matter (and perhaps even time and space) take on a grainy quality, rather than a continuous smoothness. This limit to measurement has revealed something very fundamental about the stuff we're made of—and the space we live in.

Another natural limit of measurement is the speed of light. Nothing travels faster than light, so light speed becomes the ultimate yardstick for speed in the universe. In fact, attempts to measure changes in the speed of light as it passed through the so-called luminiferous ether were some of the most successful scientific failures of all time—because they eventually helped to prove that there *was* no luminiferous ether, and also because they got Einstein, for one, to thinking about the nature of speed.

The speed of light, it turned out, is always measured as constant—always 186,000 miles per second—not only passing through the nonexistent ether in any direction, but also in situations where it should logically register changes. This is a very odd and surprising result of measurement.* The light coming at us from a star, for example, comes at us at the same speed whether we are rushing toward the star or rushing away from it—some-

*See "Relatively Speaking."

thing that makes about as much sense as having the speed of an approaching car remain the same whether you are approaching it or standing still. (Obviously, the car should approach faster if you are also approaching it.) But this does not happen with light. No matter how fast you approach a source of light or move away from it, its measured speed is always the same.

Einstein took this queer "limit" to measurement and used it as a springboard for thinking about speed. What does speed mean, anyway? Speed is only the product of time and distance. So if the speed can't change, perhaps the time and the distance can. Intuitively it seems that time and space (distance) are constants, but that speed—as an arbitrary linking of the two by human beings—can take infinite values. Einstein showed that the opposite is true. The speed of light is a universal constant. And space and time are the infinitely variable inventions of human beings—"concepts amalgamated largely from measuring devices built of matter, and not vice versa," according to Guy Murchie.

Measurement, that is, can tell us much more about something than merely *how much of it* there is. It also tells us *what* it is. Heat and temperature, for example, were considered to be one and the same thing until attempts to measure them precisely revealed that they were quite different. The temperature of the gas inside a cool fluorescent light can be ten thousand degrees, even though it contains little heat. A large chunk of earth can contain a great deal of heat, even though its temperature remains cool. This is because heat is the total amount of molecular motion in something, while temperature is the average energy of that motion. So that even though the molecules inside a fluorescent tube fly about with tremendously high energies (tremendously high temperatures), there simply aren't enough of them to add up to much heat.

In the same way, many aspects of motion remained mysterious until Newton took a more measured look and discovered that "motion" was really a combination of many different things: mass, acceleration, velocity, energy, momentum, and so on. So measurement becomes very much a means of perception. "Language . . . and measurement with instruments to check the accuracy and extend the range of the senses," writes Richard Gregory, "are the two crucial developments which distinguish man from all other animals. They have allowed us to take the extraordinary step of developing accounts of the world quite different from the way we see it."

To make a thermometer in the first place, you need to know something about how the energy of molecular motion in a roast is related to the rising column of mercury in a tube. To measure very hot things, like stars, you need to know how temperature is connected with color; to measure very cold things, you need to know how cold affects electrical resistance in various kinds of metals.

My friend the physicist has often speculated that our understanding of intelligence today corresponds in some ways to our old notions of heat and temperature and motion: we are frustrated in our attempts to measure it mainly because we have no clear idea of what it is; we treat it as a single quantity when it is probably a complicated combination of several different qualities. The fickleness of trying to put our fingers on intelligence is amply illustrated by "geniuses" like Einstein and Niels Bohr, the father of quantum mechanics. Everyone knows that Einstein was a terrible student—inattentive and bored—in school. Victor Weisskopf insists that both Bohr and Einstein would probably flunk today's intelligence tests, because they were "slow"—a term most people take to be synonymous with "stupid." George Gamow has great fun with Bohr's slowness in his *Biography of Physics:* "Perhaps his most characteristic property was the slowness of his thinking and comprehension," writes Gamow. When the physicists went to the movies, Bohr "could not follow the plot, and was constantly asking us, to the great annoyance of the rest of the audience, questions like this: 'Is that the sister of that cowboy who shot the Indian who tried to steal a herd of cattle belonging to her brother-in-law?'"

Physicist Hendrick Casimir attributes Bohr's legendary slowness to the fact that "he thought about so many sides of a question that it took him a long time to reach a conclusion." It's frightening to think that a deep and careful consideration of ideas could be considered a lack of intelligence! But no doubt it shows up as that on many standard tests. Weisskopf suspects that what most intelligence tests really measure is something closer to "smartaleckism." Once he even got annoyed when someone described his old friend Bohr as "brilliant." Someone who is brilliant, he said, "has an answer for everything. Bohr had a question for everything." Having answers and finding questions are two separate aspects of intelligence, just as speed and acceleration are two separate aspects of motion.

Thinking surely has as many facets as motion or heat. (Or as Bohr said to Einstein during their great controversy over quantum mechanics: "You are not thinking. You are merely being logical.") Which aspect of thinking do we measure as intelligence? The truth is that any attempt to measure something tends to concentrate on what is *easiest* to measure. We measure what lends itself to measurement, in the same way as we tend to judge, say, "success" by such easily definable symbols as cars, college degrees, and corner offices. Intelligence testing started out as a simple measure of things like visual and auditory acuity, strength of pull, and reaction times. It's obvious that when we consider only the measurable, we're bound to miss a lot—not only depth and creativity, but also humor, friendship, beauty, and a score of other things. It's not so obvious that the lure of measurable things may mislead us about the very nature of matter.

Which brings us back to Heisenberg. Einstein never liked the uncertainty principle because he didn't like the notion that there were things we couldn't—not even in principle—measure. (There is a curious parallel here in intelligence testing: when French psychologist Alfred Binet declared at the turn of the century that such testing was too complex to yield clear-cut answers, people complained because he was being "unscientific.") What Einstein saw as an intellectual dead end, however, Bohr and others saw as a philosophical treasure chamber, a great opportunity to look at nature in fresh ways. The answers were turning out to be limited only because the questions were probably inappropriate. "The Heisenberg relations are warning posts that say use ordinary language only up to here," says Weisskopf. "When you get to the dimensions of the atom, you get into trouble. The uncertainties come in only if you insist on trying to describe atomic states with ordinary concepts."

It's true that you can't tell the exact location and motion of an electron at the same time. But location and motion may not be concepts that make sense in the realm of the electron. The uncertainty may result mainly from a misapplication of the metaphor. As British physicist Sir James Jeans put it: "It is probably as meaningless to discuss how much room an electron takes up as it is to discuss how much room a fear, an anxiety, or an uncertainty takes up."

Or as Morrison remarked when I pressed him about certain "innate" properties of matter: "Physics doesn't talk about innate. If you can't measure it, you don't know if it's true or not. That doesn't mean that it's true or not true, it just means it's outside of the physical relationship."

No wonder properties of subtomic particles always boil down to quantities like energy, momentum, mass, and spin—because those are the things you can measure in a laboratory. All the rest is imagery and extrapolation. The measurable is definite, concrete. The unmeasurable seems arbitrary and unreal. "The immeasurable thus we naturally attribute to chance, or God," says Murchie, "and only the measurable, the clearly calculable, to science or human reason."

Taking this one not very large step further, the way we measure and therefore define something determines whether or not it exists. When we ask whether electrons are waves or particles, we are asking loaded questions. (And the answer is radically different depending on how you pose the question.) We get what we ask for. You can't just sit back and ask the "real" electron to please stand up. The answer to the question of whether electron waves exist depends on what you mean by existence, writes George Gamow.

> Wave functions "exist" in the same sense as the trajectories of material bodies. The orbits of the earth around the sun, or of the moon around the earth, *do exist* in the mathematical sense representing the continuum of points occupied consecutively by a moving material body. But they *do not exist* in the same sense as the railroad tracks which guide the motion of a train across a country.

Orbits and waves are things that can be precisely measured, but which may or may not "exist." They are relationships, like the laws of nature. If they have a reality, it is very different from the one we are used to.

In the same way, it is easy to measure many of the things that children learn in school. But whether or not those things are useful or important or even "real" in the context of the child is an open question. Unfortunately, many people think that only measurable quantities are significant. They don't consider anything worth learning unless it can be tested, or anything worth doing unless it produces a tangible reward. They may not even consider their lives (or those of others) worthwhile unless they can be measured in terms of material or social symbols. But the true richness of measurement lies in the very realization that almost everything in nature—including people—is as multifaceted as heat or electrons. There is no one way to measure anything; there is no one way to observe anything.

The true value of measurement lies elsewhere. Very simply, it can be the first step in making a complex problem soluble. As Sir Peter Medawar wrote in his *Advice to a Young Scientist:*

> Very often a solution turns on devising some means of quantifying phenomena or states that have hitherto been assessed in terms of "rather more," "rather less," or "a lot of," or—sturdiest workhorse of scientific literature—"marked" ("The injection elicited a marked reaction"). Quantification as such has no merit except insofar as it helps to solve problems. To quantify is not to be a scientist, but goodness, it does help.

Of course, the quantum leap from reality to *measurable* reality (just like the leap from reality to words) inevitably loses something in the translation—whether we are trying to put an abstract finger on human qualities or a macroscopic finger on submicroscopic quantities. Since we can't perceive quantum mechanical things directly, for example, we have to extrapolate them from everyday measuring tools like meters and clocks and images on photographic plates. Yet meters and clocks and photographs do not have quantum mechanical properties. In the same way, we are bound to lose something when we try to measure love or success or intelligence with measures that are not themselves lovely or successful or intelligent. The most important qualities of the measuring tool may not apply in the realm of what is being measured, and vice versa. In the end, insisting on measuring things with inappropriate tools may even destroy them.

I once had a friend who made careful lists of all the qualities of her friends: pluses in one column; minuses in another. On the advice of a therapist (no less), she tallied them up in order to decide who was a true friend and who was not. By the time she got through abstracting and equating, of course, she had missed whatever qualities may have been important. In the same way, it's possible to dissect a conversation or a symphony down to its last word or quarter note: whether or not you retain anything of the original in the end seems extremely doubtful.

Measurements also must be interpreted—and so are often misinterpreted. The illusion that the moon grows bigger and smaller at various times of the year and month* is really an error in measurement. And our perceptual measurements of the amount of brightness in a scene or the amount of motion we experience are vulnerable to the adaptation that makes information from the outside world "stop registering."

Moreover, the interpretation of measurements can get stuck in unconscious desires and predudices. "A scientist may have spent his life arguing that a certain combination of tooth and jaw characteristics or a certain skull size can mean only one thing," writes Johanson in *Lucy*. "When some new evidence comes along to challenge his idea, it is hard for him to accept it."

Is there any way to circumvent these various limits to measurement? Not too long ago, I spoke with psychologists Alan and Nadeen Kaufman, of the California School of Professional Psychology at San Diego, authors of a new kind of intelligence test for children. The Drs. Kaufman were well aware of all the obstacles to intelligence testing, and provided much of the history included in this chapter. I was amazed at how often obstacles to measuring intelligence paralleled obstacles to measuring the material things of physics. But I was even more amazed that the Kaufmans' exit from the maze of entanglement seemed to parallel very closely the physicists' solution: that is, two wrong approximations *can* add up to a right.†

Or to put it more precisely, the Kaufmans felt that they had devised the first intelligence test that reflects the tremendous

*See "Seeing Things."
†See "Natural Complements."

amount of new knowledge about how the brain works—knowledge
that has resulted from research over the past decades. And what
does that knowledge tell us? However you slice it (left and right,
front and back) the brain seems clearly divided into separate
areas, each with a different approach to thinking and method of
solving problems. One area specializes in logical, analytical, se-
quential (mathematical) thought, while the other specializes in in-
tegrative, holistic, creative thought. The two seem to complement
each other just as the image of light as waves complements the
image of light as particles. Intelligence, the Kaufmans have concluded, lies not in either sphere alone nor even in the two of
them together—but rather in the integration of the two, the ability to switch back and forth at appropriate times. It is much
like knowing when to switch back and forth between images in
science—between mathematics and visualization, numbers and
language, words and pictures. I don't know whether or not the
Kaufmans are right, or even whether "intelligence" is worth measuring in the first place, but I must admit the idea has a nice
ring to it.

Weisskopf begins his book *Knowledge and Wonder* with an introduction to the size of things around us: How big are we compared to stars, or atoms? How old are we compared to a cell, or a
world? It makes one marvel that people could measure such things
at all—much less count the two-hundred-billion-odd stars in our
galaxy, or clock the infinitesimally short lifetimes of quarks.
Weisskopf quotes the French philosopher Blaise Pascal who said:
"It is not the vastness of the field of stars which deserves our admiration, it is man who has measured it."

9. NATURAL COMPLEMENTS

Victor Weisskopf likes to tell the story about a conversation that took place many years ago between Nobel Prize winners Felix Bloch and Werner Heisenberg. The two famous physicists were walking along the beach, and Bloch was sounding out Heisenberg—so the story goes—on the significance of some new theories having to do with the mathematical structure of space. At length, Heisenberg responded, "Space is blue and birds fly in it."

The story is especially nice because it illustrates so well what many physicists believe to be the most profound contribution of

quantum theory. This contribution was not a discovery in the normal sense: not a particle, not a new kind of extraterrestrial object
or event, not even a theory or equation. It was, rather, a philosophical outlook that allowed scientists to see beyond the mass of
paradoxes that seemed to be making modern physics all but impenetrable. It was the notion of complementarity. And if nothing else,
the story of Heisenberg and Bloch captures the essence of complementarity: that one can talk about the same subject in two very
different kinds of terms, and that what makes good sense in one
context can make absolutely no sense in another.

Complementary ideas are opposing ideas that add up to much
more than the sum of their parts. They complement each other like
night and day, male and female, "on the one hand" and "on the
other hand." Complements are required for a full spectrum of understanding, just as a full array of colors is required to produce
pure white. (In fact, any two colors are considered complementary
if they add up to white.) Complements are the yin and yang of
science. Or as the physicist Emilio Segrè wrote, "It is one of the
special beauties of science that points of view which seem diamet-
rically opposed turn out later, in a broader perspective, to be both
right."

Complementarity is almost everyone's favorite when it comes
to "sentimental fruits of science." And no wonder: life, like nature,
is fairly bursting with unresolved—unresolvable—paradoxes.
Particles are waves, and waves are particles. You are ninety-eight
cents' worth of cosmic star dust floating at the obscure edge of an
ordinary galaxy, and yet you are the center of your own world; to
friends and family, you may be precious beyond all worth. On one
day, humanity seems the apex of all things beautiful, generous,
mindful; another day, it seems a stupid beast. But the truth is that
we are both beauty *and* the beast—just as energy is only another
way to look at matter.

Danish physicist Niels Bohr fathered complementarity as a
way to tame the inherent contradictions in the Heisenberg relations: the fact that you could not pinpoint the position of a particle
without sacrificing knowledge of its motion, or measure its motion
without introducing uncertainty about its position; and also the
paradox inherent in the discovery that all stable forms of matter in
the universe are fashioned from indivisible bits of vibrating energy

known as quanta—and yet if you try to probe such a quantum to see how it is made, it will melt as surely as a snowflake in the palm of your hand.

Bohr said that the reality of particles required *complementary* descriptions, more than one point of view. It doesn't matter that you can't measure both motion and position at the same time; you can't see both sides of a coin at the same time either. As long as people insist on viewing the subatomic world with their everyday perspectives (and what choice have they?) they will be stuck with looking at nature one dimension at a time. "In our description of nature," said Bohr, "the purpose is not to disclose the real essence of phenomena but only to track down, as far as it is possible, relations between the manifold aspects of experience."

The idea of complementarity reaches far beyond Heisenberg. It also helps to explain the "wave/particle duality" of light. By the turn of the century, decades of argument and experimentation had finally convinced physicists that light had to be a wave. It bent, or diffracted, around corners as an ocean wave bends around a reef. Two sets of light beams produced the bright and dark interference bands that result when waves reinforce and then cancel each other. Then Einstein came along and showed that light came in parcel-like clumps. Light was "quantized." Things seemed so contradictory that one physicist was moved to remark that nature behaved according to quantum theory on Mondays, Wednesdays, and Fridays, and wave theory on Tuesdays, Thursdays, and Saturdays.

It seemed obvious to everyone that light could not be *both* waves and particles. The notion of a wave is as different from the notion of a particle "as is the motion of waves on a lake from that of a school of fish swimming in the same direction," says Weisskopf. A particle is like a bullet—a patently material, finite object that occupies a particular place in space and time. A wave is more like a motion—a continuous abstract form. If the wave theory was right, then the particle theory had to be wrong. If one was truth, the other was by definition heresy.

It turns out, of course, that both are right. Waves and particles are complementary ways of describing the nature of light—just as position and motion are complementary aspects of particles. Not only light, but also all energy and matter and radiation reveal the same curious duality. Electrons can be diffracted like light beams, and show the same interference patterns when passed through the layers of aligned molecules in a crystal.

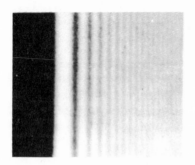 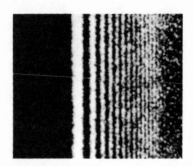

Interference patterns produced by light waves (left) are almost identical to those produced by beams of electrons (right). Both radiation and matter have a curious duality: sometimes they behave like waves, at other times like particles.

The thing that's hard to swallow about complementarity is that the various aspects of a situation needed to make it "whole" can also be mutually exclusive—like waves and particles. Complements are more than a fancy physicist's version of "on the one hand/on the other hand." When you look at one side of a coin, you may not be able to see the other side of the coin—but the unseen side does not seem impossible or absurd. Just as when you look at the familiar illusion of "vase and faces," only one image or aspect is visible at a time—but that doesn't rule out the possibility of the other point of view.

Waves and particles, on the other hand, seem to present a much clearer choice. They do not make sense in the same context, and it is hard to see how both can describe the same thing. In the same way, considering yourself as the pivot of your universe seems to require that you abandon the idea of people as ultimately "ashes and dust"—and vice versa.

The trick to complementarity is knowing when which view is appropriate. For as J. Robert Oppenheimer pointed out, "The more nearly appropriate the first way of thinking is to a situation, the more wholly inappropriate is the second."

You cannot see both the profiles and the goblet at the same time; you must continually shift your point of view.

The statement that "space is blue" is hardly an appropriate way of expressing a mathematical relationship—but then an equation would hardly be an adequate way of describing one's sense of the sky during a stroll on the beach on a summer's day. The more a particle acts like a particle, the less it behaves like a wave—just as the clearer the position of a particle becomes, the more fuzzy its motion becomes (and vice versa).

(The term "complementarity" comes, of course, from colors. Blue and orange are complementary because they add up to white —the full spectrum. They are also mutually exclusive. Blue is a color that contains no orange, just as orange contains no blue. If you subtract blue from white, the leftover light is orange. A complement is a kind of a shadow. Night is both the shadow, and the complement, of day.)

In the most extreme cases, focusing sharply on one aspect of a situation actually destroys the other. Bohr must have been a skier, for he put the paradox this way: "When you try to analyze a christiana turn into all its detailed movements, it will evanesce and become an ordinary stem turn, just as the quantum state turns into classical motion when analyzed by sharp observation." Weisskopf, the pianist, compares it to Beethoven: "We cannot at the same time experience the artistic content of a Beethoven sonata and also worry about the neurophysiological processes in our brains. But we can shift from one to the other."

A quantum state, like a ballerina's arabesque or a bird's song, retains its character only as long as it remains whole. This is not to say that you cannot break down these things into their individual notes and motions and molecules—only that watching an animal in the wild and dissecting it in a laboratory are complementary ways of exploring nature.

These "manifold aspects of experience," as Bohr called them, can be mutually exclusive not only because you destroy one in the process of observing the other, but also because what you see depends on the kind of observation you make. That is, light shows up as a wave in experiments designed to detect interference patterns. But it shows itself as a particle in other experiments that measure the way an atom emits its energy as light.

In this sense, particles share some attributes of people: they start off with the potential to take many forms. We cannot say that they are innately waves or particles, passive or aggressive, good or bad. What they become depends largely on the obstacles and opportunities we place in their path.

This is not to say that how people behave is governed entirely by outside circumstances. Nature and nurture are two sides of the human coin—complementary aspects of the argument. In the same way, recognizing our biological ties with the animal kingdom does not relegate us to a life as beasts, stepping all over each other's rights and feelings and lives in the struggle for "survival of the fittest." As Loren Eiseley points out, early humans were small in stature and scant in numbers, so that cooperation must have been at least as important to our success as a species as competition. If "eat or be eaten" seems an essential ingredient in the jungles of today's life, then so are honesty and generosity. We make both love *and* war. "Violence, sexism, and general nastiness *are* biological since they represent one subset of a possible range of behaviors," writes Stephen Jay Gould. "But peacefulness, equality, and kindness are just as biological—and we may see their influence increase if we can create social structures that permit them to flourish."

This human duality is often as hard to accept as the physical duality of wave and particle. And it can be dizzying to follow the swing of the pendulum from one philosophical or political fashion to a contradictory one. We seem to be going around in circles, never making any human progress. "In one period we believe ourselves governed by immutable laws; in the next, by chance," writes Eiseley.

> In one period angels hover over our birth; in the following time we are planetary waifs, the product of a meaningless and ever-altering chemistry. We exchange halos in one era for fangs in the next. Our religious and philosophical conceptions change so rapidly that the theological and moral exhortations of one decade become the wastepaper of the next epoch. The ideas for which millions yielded up their lives produce only bored yawns in a later generation.

He then quotes Montaigne: "We are, I know not how, double in ourselves, so that what we believe we disbelieve, and cannot rid ourselves of what we condemn."

As unnerving as this duality may be, however, it may also be the essence of our survival—the multiplicity of forms that is embodied in everything from species to democracy, the ability to switch gears that provides both resiliency and strength. It means that any view of nature or human nature that views one "side" as dogma and the other as heresy is probably wrong, or at least dangerous. As Weisskopf points out, this domination of one idea has inevitably led to abuses—whether it was the dominance of religious dogma during the Middle Ages, or the excessive influence of technology today. "Whenever one way of thinking has been developed with force, claiming to encompass all human behavior, other ways of thinking have been neglected. This has its roots in a strong human desire for clear-cut, universally valid principles containing answers to every question. But because human problems always have more than one aspect, general-purpose answers do not exist."

The answers are not either/or but "all of the above"—or at least "each" of the above at times when they are appropriate.

M.I.T. professor of computer science Joseph Weizenbaum said much the same thing recently in speaking about, of all things, the dangers of overreliance on computers. Weizenbaum, who wrote *Computer Power and Human Reason,* is afraid that society's love affair with computers is a symptom that the scientific mode of thinking is becoming "imperialistic." Not that scientific thinking is bad—only that it becomes dangerous when it overwhelms all other approaches. "If you wanted to understand the Great Depression of the 1930s and you only looked at Department of Labor statistics and you didn't read novels by people like John Dos Passos because novels are not scientific, then that's bad—because in a very deep way you can learn more by reading the novels."

Weizenbaum's worry is hardly new. Nobel Prize winner Max Born carried on a long and well-known correspondence with Albert Einstein over several of the perplexing implications of the new physics—specifically whether the notion of cause and effect could survive quantum theory, or whether, as Einstein refused to concede, it meant that "God plays dice with the universe."* Born concluded that their argument was at root a reflection of the unfortunate inclination of philosophy and also science toward "final, categorical statements." Order and chance only *seem* to be mutually exclusive to minds accustomed to either/or alternatives. The laws of probability are *laws* of nature, too. Born wrote: "If quantum theory has any philosophical importance at all, it lies in the fact that it demonstrates the necessity of dual aspects and complementary considerations. . . . Much futile controversy could be avoided in this way."

Stretching the metaphor still further (even, admittedly, perhaps out of bounds), the notion of complements makes it easier to reconcile many of the either/or arguments that leave society bound in so many unpleasant knots: concern for the unborn *or* legal abortions; a strong national defense *or* efforts toward peace; tightening the bureaucratic belt *or* help for the poor; isolationism *or* interventionism; women's rights *or* concern for the family and children; pornography *or* censorship. And so on. I've even heard people in-

*See "Cause and Effect."

sist that The Exploratorium is a "children's museum" merely because the people inside are so clearly having fun, and because everyone knows that "real" science is hard and boring. Real science does, of course, require study and practice and precision and lots of hard work—just like painting and baseball and business and running a home. Everything human beings do, for that matter, requires aspects of work and aspects of fun, aspects of logic and aspects of fancy, aspects of the trivial and aspects of the profound.

For centuries, people argued over whether light was essentially a wave or essentially a particle. Today, this seems as superfluous as arguing about whether space is blue or whether it has mathematical properties. Each, in its proper context, is true. This doesn't imply that the "whole truth" lies somewhere in between the two viewpoints: complementarity is not a compromise. It is rather like the sides of a box, or the facets of a problem. What you see depends on what side of the box you look at—which is why light and in fact all energy and matter show up as clumps or quanta in some experiments, and behave like waves in others.

As such, complementarity makes it easier to accept the innate limits on perception, measurement, and the extent to which we can imagine the unseeable world.* Each way of seeing goes only so far. Like the image engraved on the back of each eyeball by incoming light beams, all models and perceptions are flattened representations of reality. They are projections that lose at least one dimension in the process of imprinting themselves in our minds. And just as we need two eyes to see depth—the combination of two distinct images—so we need more than one perspective to see almost anything in all its dimensions.

Even a microscope enhances your ability to see only at the expense of the larger context. If you put a living organism under a microscope, you can see far more clearly into its individual membranes and cells. But even then, as Weizenbaum points out, "it wouldn't make sense to say that what you're seeing in any way resembles the essence of the organism itself." If you need to see the individual trees, you will necessarily lose the view of the larger forest; yet if you opt to see the forest you lose sight of the details of any individual tree.

*See "Seeing Things," "Science as Metaphor," and "The Measured Approach."

Accepting complementarity merely means accepting that because one view is right, the opposite view isn't necessarily wrong
—that truth is not the other side of heresy (or vice versa); that if
you look at a problem "scientifically," considering its moral aspects
isn't "besides the point"; or that considering a course of action
based on emotion or ethics doesn't preclude learning a little more
about the situation "scientifically." If people are somehow coming
to the conclusion that science is a one-way street to a single right
answer, that would be profoundly ironic, for as Born wrote: "This
loosening of the rules of thinking seems to me the greatest blessing
which modern science has given us. . . . The belief that there is
only one truth and that oneself is in possession of it seems to me
the deepest root of all the evil that is in the world."

10. THE SCIENTIFIC AESTHETIC

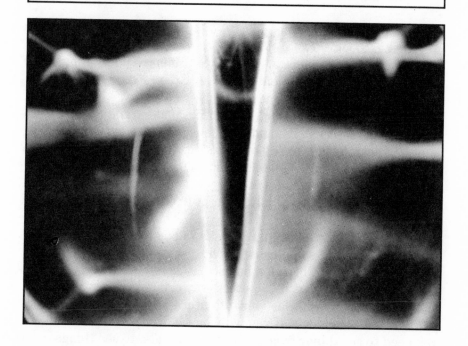

Poets say science takes away from the beauty of the stars—mere globs of gas atoms. Nothing is "mere." I too can see the stars on a desert night, and feel them. But do I see less or more? The vastness of the heavens stretches my imagination—stuck on this carousel my little eye can catch one-million-year-old light. A vast pattern—of which I am a part—perhaps my stuff was belched from some forgotten star, as one is belching there. Or see them with the greater eye of Palomar, rushing all apart from some common starting point when they were perhaps together. What is the pattern, or the meaning, or the *why*? It does not do harm to the mystery to know a little about it. For far more marvelous is the truth than any artists of the past imagined! Why do the poets of the present not speak of it? What men are poets who can speak of Jupiter if he were like a man, but if he is an immense spinning sphere of methane and ammonia must be silent?

This poetic paragraph appears, of all places, as a footnote in a physics textbook—*The Feynman Lectures on Physics*, to be exact. It is a most eloquent argument that science, if anything, *adds* to an aesthetic appreciation of nature. Science does not strip nature bare of its emotion and beauty, leaving it as a naked set of equations. On the contrary, a scientific understanding of nature only deepens the awe, expands the sense of mystery. Or as Weisskopf has been known to say on occasion: "I enjoy lying in the sun and feeling the result of the nuclear reactions. I don't enjoy it any less because I understand it; I enjoy it more."

Science and art are usually thought to be as mutually exclusive as waves and particles. And so they are. That is, they are a prime example of how complementary ideas can so nicely, well, complement each other. As one artist put it, art provides the peripheral vision we needed to broaden our scientific eye on things—but he might as well have put it the other way around. For both science and art select a particular view of nature and bring it into focus, put a "frame" around it, if you will. These frames are the windows you need to see things from more than one side, to see things in all their perspectives. The many aspects of art and science are so intricately intertwined that one is often hard put to answer, Is it science? Or is it art?

This is especially true of many of the pieces built by artists at The Exploratorium. Not so long ago, I was talking with one artist about an "exhibit" he had built called Quiet Lightning—a sparkly pink ball of glowing electric charge. I asked if he had built it primarily as a sculpture or as a demonstration of a scientific idea. In other words, was it science, or was it art? He thought for a while and then said: "It's neither. It's nature." Nature, that is, modified by the multitude of decisions made in the process of looking, seeing, recording.

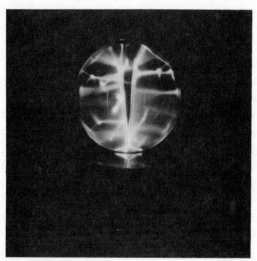

Bill Parker's Quiet Lightning: is it science, or is it art?

Many social phenomena similarly defy categorization. Is teaching a science, for example, or is it an art? How much art is there to political science, or to the "science" of medicine? We talk about the art of conversation, and how-to books promote what they call the science of success. History, music, architecture, salesmanship, and a million other things are both sciences *and* art. Even love: it is Cupid's potion and the province of poets, but it also can be described in terms of electrical charges—calculated chemical reactions contrived to ensure the survival of the species.

Of course, when people talk about a dichotomy between science and art they usually are talking about the approach, not about the subject matter. Both a poet and a botanist might consider a tree as fertile ground for investigation. The relationship between mother and child, the symmetry of snowflakes, the effects of light and color, and the structure of the human form are studied equally by painters and psychologists, sculptors and physicians. The origins of the universe, the nature of life, and the meaning of death are subjects explored by physicists, philosophers, and composers.

Yet when it comes to approach, the affinity breaks down completely. Artists approach nature with feeling; scientists rely on logic. Art elicits emotion; science makes sense. Art, like childrearing or an interest in social welfare, is supposed to require a warm (if not bleeding) heart. Science, like law or manufacturing, is supposed to be rational, objective, deductive. Scientists are supposed to think, but artists are supposed to care.

Or so goes the standard view of things. In truth, nothing could be further from the truth—for it would be a lousy poet indeed who was "uncontrolled," and no scientist ever got very far by sticking exclusively to the "scientific method." In fact, although for some reason it seems to upset people, scientists are frequently every bit as passionate about their work as artists are about theirs.

Take Charles Darwin, for example. While he was rummaging around the Galapagos Islands gathering the evidence that would eventually lead to his theory of evolution and natural selection, he was hardly what you'd call detached. "I am like a gambler and love a wild experiment," he wrote. "I am horribly afraid." "I trust to a sort of instinct and God knows can seldom give any reason for my remarks." "All nature is perverse and will not do as I wish it. I wish I had my old barnacles to work at, and nothing new." And so on.

Such passion was also rampant in the early days of the quantum debate. Albert Einstein said that if classical notions of cause and effect had to be renounced, he would rather be a cobbler or even work in a gambling casino than be a physicist. He based his objections on what he called "an inner voice. . . . The theory accomplishes a lot, but it does not bring us closer to the secrets of the Old One." Danish physicist Niels Bohr called Einstein's attitude "appalling," and accused him of "high treason." Another major contributor to quantum theory, Erwin Schrödinger, reportedly said: "If one has to stick to this damned quantum jumping, then I regret having ever been involved in this thing." On a more positive note, P.A.M. Dirac talked about his discovery of the equation that eventually led to the discovery of antimatter as "the most exciting moment of my life." And Einstein talked about the universe as a "great, eternal riddle"that "beckoned like a liberation." Hardly the stuff of your stereotypical "dispassionate" scientist. And these are but a few examples. Ernest Lawrence literally danced around the room when his assistants managed to push his newborn cyclotron to one million volts. Experimental physicist Vera Kistiakowsky once compared the excitement of what she does with "being the first person on the moon."

"When people say science is objective, unlike the humanities and arts, they are right in that scientists are careful to get their numbers right and careful not to put their prejudices into the results of experiments," she says. "But science is in fact a humanity. The only way you can understand science is through the human mind with machinery built by human beings and theories devised by human beings. So science is a human endeavor." Or as George Sarton put it in his *History of Science:* "There are blood and tears in geometry as well as in art. . . . It would be very foolish to claim that a good poem or a beautiful statue is more humanistic or more inspiring than a scientific discovery; it all depends upon the relation obtaining between them and you."

If physicists have taken to giving new properties of matter names like "beauty" and "truth" and "charm," it is not because they find one kind of matter more beautiful or truthful or charming than the next. It is rather because in some way they are trying to convey a sense of the emotion and mystery (and humor!) in what they do.

Yet somehow the idea that scientists are supposed to be *objective* about their work has come to include the idea that they are not supposed to *care* about it. "What a strange misconception has been taught to people," says my friend the physicist. "They have been taught that one cannot be disciplined enough to discover the truth unless one is indifferent to it. Actually, there is no point in looking for the truth unless what it is makes a difference."

This misconception extends far beyond physics. The fact that women seem more sensitive to issues such as peace, children, and environmental safety than men is often used as an argument that they are also necessarily softheaded about the same issues—and therefore incapable of adding anything to the argument. On the other hand, I am constantly running into student writers who don't want to hear how much discipline, rote, and practice goes into the "creative" work they want to do. We dismiss what is emotional as "untrue" or "unimportant" or even "unscientific"; yet we insist that what is precise can't be creative. Not too long ago, I heard a politician being soundly routed for the crime of getting hot under the collar—and I remembered what happened to Bella Abzug (who shouted) and Ed Muskie (who cried). Emotion seems unallowable in politics, yet human dreams and hopes and fears are what politics is all about. If we don't care deeply about the issues, why even bother to play the game?

Great artists and great scientists are often known for combining *both* approaches in their work. (And I won't even mention Leonardo da Vinci.) Artists need a scientific (or at least technical) knowledge of the materials they use—paints, paper, marble, lenses, strings, computers, and so on. A musician may know as much as a physicist about resonance, acoustics, and harmony; a photographer, as much about the optical properties of light. Even the content of the artist's work is more often than not based on the scientist's perception of nature; after all, it was not so long ago that science was "natural philosophy." Art, like all forms of perception, is conceived in a cultural context, and much of that context is construed from the perspectives people have on their physical world.

Scientists, for their part, rely on an artistic approach for those all-important leaps of the imagination called insight. Without it, they would be forever stuck in their past perceptions. "Facts do not 'speak for themselves,'" writes Stephen Jay Gould. "They are read in the light of theory. Creative thought, in science as much as in the arts, is the motor of changing opinion." Science alone logically leads us smack into our perceptual limitations, our inability to imagine (much less recognize) the unknown. "This is the power of art," writes Richard Gregory—pictures can "make us see things differently by changing our object-hypothesis." They can expand our reservoirs of possible realities. The artistic perspective is the fresh outlook that breaks established routines, that sees through all the old obvious notions and prejudices.

"Each time we get into a 'log-jam,'" writes Feynman, "it is because the methods that we are using are just like the ones we have used before. The next scheme, the new discovery, is going to be made in a completely different way. So history does not help us much." Deduction only takes you to the next step in a straight line of thought, which in science is often a dead end. As Feynman concludes, "A new idea is extremely difficult to think of. It takes fantastic imagination."

The direction of the next great leap of imagination is guided as often as not by the scientist's vision of beauty. Einstein's highest praise for a theory was not that it was good, but that it was beautiful. His strongest criticism was, "Oh, how ugly!" He often spoke and wrote about the *aesthetic* appeal of ideas. "Pure logic could never lead us to anything but tautologies," wrote the French physicist Henri Poincaré. "It could create nothing new; not from it alone can any science issue."

Poincaré describes the role that aesthetics plays in science as that of "a delicate sieve," an arbiter between the telling and the misleading, the signals and the distractions. Science is not a book of lists. The facts need to be woven into theories like tapestries out of so many tenuous threads. Who knows when (and how) the right connections have been made? Sometimes, the most useful standard is aesthetic. Schrödinger refrained from publishing the first version of his now-famous wave equations because they did not fit the then-known facts. "I think there is a moral to this story," Dirac commented later, in a now-famous quote. "Namely, that it is more important to have beauty in one's equations than to have them fit experiments. . . . It seems that if one is working from the point of view of getting beauty in one's equations, and if one has a really sound insight, one is on a sure line of progress."

It is the aesthetic sense that prompts scientists and other people as well to insist that an idea "looks" right, or "feels" wrong.

Sometimes, the connection between art and science can be even more direct. Niels Bohr was known for his fascination with Cubism —especially "that an object could be several things, could change, could be seen as a face, a limb, a fruit bowl," as a friend of his later put it. Bohr went on to develop his philosophy of complementarity, which showed how an electron could change, could be seen as a wave, a particle. Like Cubism, complementarity allowed contradictory views to coexist in the same natural frame.

Some people wonder how art and science ever got so far separated in the first place. The image of Einstein with his violin is almost as familiar as the image of Leonardo with his inventions. It is a standing joke in some circles that all it takes to make a string quartet is four mathematicians sitting in the same room. Even Feynman plays the bongo drums. (Although he finds it curious that while he is almost always identified as the physicist who plays the bongo drums, the few times that he has been asked to play the drums, "the introducer never seems to find it necessary to mention that I also do theoretical physics"—something he attributes to the fact that people may respect the arts more than sciences.) Still, Feynman has also been known to remark that the only quality art and theoretical physics have in common is the joyful anticipation that artists and physicists alike feel when they contemplate a blank piece of paper.

In truth, the definitions of both art and science have narrowed considerably since the days when science was natural philosophy and art included the work of artisans like those who build today's fantastic particle accelerators—"the cathedrals of contemporary civilization," as some have called them. "Why not aesthetics in science?" asks social theorist Sir Geoffrey Vickers in Judith Wechsler's excellent collection of essays *On Aesthetics in Science*.

> Whence comes the implication that to find aesthetics in science is like finding poetry in a timetable? The answer lies in the sad history of Western culture which, over the last two centuries, has so narrowed the concepts of both Science and Art so as to leave them diminished and incommensurable rivals—the one an island in the sea of knowledge not certified as science; the other an island in the sea of skill not certified as art.

Before this unnatural separation, says Vickers, everyone knew that knowing was an art, that understanding required skill, that both art and science were process as well as product. Science was restricted to its present narrow meaning barely before the nineteenth century, when it came to apply to "a method of testing hypotheses," based on experiments. Conducting experiments to test theories has little to do with art. But Vickers suspects that the difference is deeper. People want to believe that science is a rational process, that it is describable. Intuition is not describable, and should therefore be relegated to a place outside the realm of science. "Because our culture has somehow generated the unsupported and improbable belief that everything real must be fully describable, it is unwilling to acknowledge the existence of intuition."

There are, of course, substantial differences between art and science. Science is written in the universal language of mathematics; it is, far more than art, a shared perception of the world. Scientific insights can be tested by the good old scientific method. And scientists have to try to be dispassionate about the conduct of their work—at least to the extent that their passions do not disrupt the outcome of experiments. Sometimes, inevitably, they do disrupt them. "Great thinkers are never passive before the facts," says Gould. "Hence, great thinkers also make great errors."

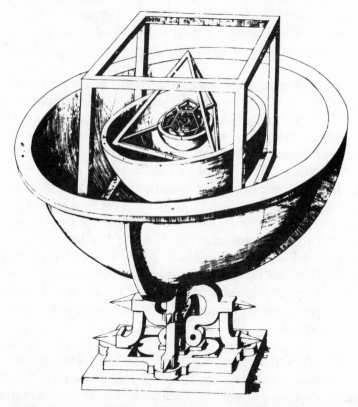

Kepler Device. Kepler's model of the universe, showing how he thought all the planets were positioned with relation to certain geometric shapes. Redrawn from Mysterium Cosmographicum *(1597, edition of 1620).*

The history of science seems to be written in these errors—primarily because the discoverers of the workings of the physical

world have been propelled as much by emotion as by thought, by
conviction as by deduction. Johannes Kepler eventually proved
that the orbits of the planets were elliptical and not round as everyone had thought—thereby laying the ground for Newton and
Einstein—but Kepler never abandoned his belief in the notion that
the orbits of the seven known "planets" were enclosed in the five
perfect geometrical solids. The way he thought about astronomy is
hardly what you'd call logical; in fact, he went so far as to argue
that the orbits that lie *outside* the earth's orbit "have it in their
nature to stand upright," whereas those *inside* the earth's orbit
have it in their nature to float. "For," he explained, "if the latter
are made to stand on one of their sides, the former on one of their
corners, then in both cases the eye shies from the ugliness of such a
sight."

This is not cited so that we modern-day scientific types may sit
back and laugh at Kepler. On the contrary, it is offered as an anti-
dote to the commonly held (and commonly wrong) assumption that
knowledge develops in a straight line of thought, conceived by
those suprahuman automatons we call scientists. Without its "irrational," aesthetic side, science would never be able to make great
leaps. On the other hand, the same nonrational side also may guide
those leaps in strange directions. This is nothing to tear your hair
about. ("Why should we allow artists, conquerors and statesmen to
be guided by irrational motives, but not the heroes of science?"
asks Arthur Koestler.) But it does imply a need for a certain caution: "Within the foreseeable future," Koestler concludes, "man
will either destroy himself or take off for the stars. It is doubtful
whether reasoned argument will play any significant part in the
ultimate decision."

This leaves the question: if scientists are as often as not guided
by a certain aesthetic, by their notion of beauty, what does
"beauty" mean to a scientist? Obviously, it does not mean pretty or
pleasing or even inspiring in the normal sense. An equation or a
theory could not be considered a thing of beauty as long as we stick
to this rather narrow definition. It turns out that beauty in science
is something much closer to the notion of *simplicity*.

"You can recognize truth by its beauty and simplicity," writes
Feynman.

When you get it right, it is obvious that it is right—at least if you have any experience—because usually what happens is that more comes out than goes in. . . . The inexperienced, the crackpots, and people like that, make guesses that are simple, but you can immediately see that they are wrong, so that does not count. Others, the inexperienced students, make guesses that are very complicated, and it sort of looks as if it is all right, but I know it is not true because the truth always turns out to be simpler than you thought.

Simplicity is beautiful to a scientist because it is a surprisingly persuasive aspect of nature, and has a great power to lead to important ideas. How many things that seemed so wildly disparate in the past have turned out to have common threads, intimate connections? The forces of nature, electricity and magnetism, falling apples and orbiting planets. As poet Muriel Rukeyser observed, even islands are connected underneath.

"It is simple, and therefore it is beautiful," says Feynman of gravity.

It is simple in its pattern. I do not mean it is simple in its action—the motions of the various planets and the perturbations of one on the other can be quite complicated to work out, and to follow how all those stars in a globular cluster move is quite beyond our ability. It is complicated in its actions, but the basic pattern or system beneath the whole thing is simple. That is common to all our laws; they all turn out to be simple things.

Of course, as Weisskopf likes to point out, what's simple is not always a matter of objective assessment. What's *simple* is what's *understood*. What you don't understand always seems complicated —no matter how simple it may seem to someone who does understand it. This explains how scientists can insist that Einstein's equations about gravity, for example, are simpler than Newton's, even though any high school student can remember and understand Newton's laws, whereas Einstein's theories require a much greater degree of mathematical sophistication. To a physicist or mathematician, Einstein's equations are broader, clearer, more logical, and therefore *simpler.*

The simplicity that so often seems inherent in nature, Sir James Jeans points out, "is of the kind which *our minds* adjudge to be simple. Indeed, any other kind of simplicity would probably escape our notice."

The simplicity, in other words, comes from the clarity of understanding, from the ability to see through the distractions and focus in on the essential elements, to explain many seemingly unconnected things with one "simple" idea. And so simplicity, beauty, and knowledge all are linked together. No wonder physicist Enrico Fermi threw up his hands at the great numbers of elementary particles appearing in the 1950s: "If I could remember the names of all these particles I would have been a botanist." Another physicist was known to grumble upon learning of the discovery of a new particle: "Who ordered this?"

Nature seems to be built on patterns, and looking for those patterns is the primary preoccupation of artists and scientists alike. "What's beautiful in science is that same thing that's beautiful in Beethoven," says Weisskopf. "There's a fog of events and suddenly you see a connection. It expresses a complex of human concerns that goes deeply to you, that connects things that were always in you that were never put together before."

The scientific aesthetic is based on the pleasure of understanding. And just as the most profound scientific insights are based on imagination and inspiration rooted in a firm foundation of fact, so artistic inspirations often come from a detailed understanding of nature. So there is nothing contradictory about putting aesthetics into our cities, humanity into our businesses, practicality into our social programs, or rigid standards into the most creative enterprises. Perhaps if we could better use science to cultivate our idealism, we could also learn to use art to guide our practical decisions.

In the end, the connections between science and art may rest on the matter of motivation. That is, both scientists and artists are seekers; they tend to share a knack for noticing and interpreting either (or both) natural or human phenomena. When M.I.T. metallurgist Cyril Stanley Smith became interested in the history of his field, he was surprised to find that the earliest knowledge about metals and their properties was provided by objects in art museums. "Slowly I came to see that this was not a coincidence but a consequence of the very nature of discovery, for discovery derives from aesthetically motivated curiosity and is rarely a result of practical purposefulness."

11. SHADOWS
AND
SYMMETRIES

Bob Miller of San Francisco is an artist who has created a whole series of sculptures at The Exploratorium on the subject of light. But far more than art or even science, the essence of his work is "natural philosophy." For his work with shadows has shown that there is as much information in the shape of the darkness as there is in the light, and his reflections reveal that the images we see often take circuitous paths to reach us—and frequently return as echoes of ourselves.

Everyone Is You and Me: *your reflection mingles with images seen through a window so completely that you cannot tell which is which.*

One of his pieces is a "window-mirror" aptly titled *Everyone Is You and Me*. It is so enchanting and intriguing that it was selected as the only American exhibit to be shown at a recent worldwide museum conference in Moscow. Yet it is nothing more than a piece of half-silvered glass that partially reflects and partially transmits light. If you sit in front of it, you can see yourself in the "mirror." But you can also look through the "window" at a person on the other side. The curious conglomeration of features that appears in the window-mirror may be an accurate reflection of the way we perceive our everyday world.

I know a woman, for example, who insisted that the average annual income in the United States must be "at least" fifty thousand dollars; of course, all she was seeing was her own image as reflected in her closed circle of friends. It reminded me of the Reagan aide who quit because (he said) a family "cannot live" on

$62,000 a year in Washington, or the young lawyer who told the *Washington Monthly* that his $370,000 house was "not that expensive." If these examples seem outrageous, they are really not so

A reflection appears behind *the looking glass, as Alice knew. In this exhibit, the reflected image of a person sitting on one side of the glass appears in the same place as the real person sitting on the other side. That is why the two images can combine.*

different from the feelings most of us get from time to time when we watch the six o'clock news: we inevitably wonder, why is it, if there are so many Republicans (or Democrats, or feminists, or Right-to-Lifers, or hungry children, or unemployed, or families with computers, or cocaine sniffers) in the world, that I don't ever seem to meet any? It's like the woman at *The Washington Post* who wondered how it was that all the media people were caught off guard when the Republicans were swept into power in 1980 in a landslide. The reason, of course, was that all the New York and Washington media people had been listening and talking mainly to each other. We had been seeing a reflection and assuming it was a window to the outside world.

Things have not changed so much since the days when the Ionian philosopher Xenophanes reflected: "Men imagine gods to be born, and to have clothes and voices and shapes like theirs. . . . Yea, the gods of the Ethiopians are black and flat-nosed, the gods of the Thracians are red-haired and blue-eyed."

Light beams are engraved with information which they carry around with them wherever they go.

Once, while surveying, Thoreau encountered an unusual echo. "After days with humdrum companions," as Loren Eiseley tells it,

Thoreau recorded with surprise and pleasure this generosity in nature. He wanted to linger and call all day to the air, to some voice akin to his own. There needs must be some actual doubleness like this in nature, he reiterates, "for if the voices which we commonly hear were all that we ever heard, what then? Echoes . . . are the only kindred voices that I hear."

You don't have to be Narcissus (or even Thoreau) to be seduced by reflections. They are as real in their own way as the so-called concrete objects around us. You don't know where a light beam has been before it gets to you even when its source seems to be right in front of your eyes. You don't know that the bright red spot of light hovering outside your car window is really a reflection of a bicycle taillight behind you, so you might see it as a UFO. Reflections take light and all its colorful images and spin them around as if in a revolving door. But the revolving door is invisible; you have no way of knowing how many times it's been spun around before it

The pointy spokes around stars and streetlights are created inside your eye as the light bends around layers of cells in the lens. A camera has to use a diffraction grating to produce the same effect.

gets to you. As I look out my window, I think I'm seeing the edges of a pond about a half mile away. But actually I'm seeing the reflection of sunlight on the pond—a reflection which has been absorbed and reemited many times by air molecules and then bent and unbent as it passed through the window, and then again by the corneal covering of my eye. I can't see the light from the pond without also seeing all that has happened to it between there and here.

All of the information and ideas we deal with are filtered through many layers of perceptions and prejudices—our own and those of people around us. This is also true of images borne on light. My son recently observed that the difference between stars and planets was that stars are "pointy." But the broad spokes of light that fan out from stars and streetlights and Christmas tree bulbs are not produced by the stars at all. A star is a great round ball of gas. (The sun is a star.) The spokes or "points" are created as the light bends around onionlike layers of transparent cells that make up the lens in your eye. The stars don't really twinkle. They only seem to shimmer because the air between us and the stars is constantly moving, shaking the light from the star around. The sparkle that we see when we look at the stars is a twinkle in the mind's eye.

As light bends through air, or water, or glass, it takes its images with it, creating all manner of "mirages."

The everyday world is replete with similar phenomena. When the light coming from the objects near a gas pump passes through the vapors rising from the hose as it fills a car, the light beams skid around, taking their images with them, making the objects behind the shimmering gas appear to shimmer and change shape, too. A toothbrush propped in a glass of water appears to bend because the light from the toothbrush bends as it passes through the water—and the properties of the water become superimposed on the image of the brush. Mirages such as the flattened "pumpkin" sun that appears at sunset are created by bent beams of light that make images appear out of place. A lens bends light by exactly the same process, although in a much more regular fashion. When you put on your glasses, you may see things more clearly, but in reality you are seeing a mirage.

Many exhibits at The Exploratorium were designed to show how the play of light can wreak havoc with our notion of what's what and what's where. Glass rods disappear when dipped in a special transparent solution; a solid spring turns into a ghost as you try to touch it; myriad kaleidoscopes make a crowd of three into a party of dozens.

Three's a crowd.

But other kinds of media that transport other kinds of messages can be equally distorting. Our friends and relatives, the press and entertainment industry, schools and churches, all refract the information they transmit in ways that reflect their own natures. If most of the "national" news we see on TV every night happens in Washington, it is not because most of the news of the nation happens in Washington; it is because most of the national *press* is in Washington. If the surgeon is always suggesting surgery, and the housewife insists that the place for women is in the home, and the military leader sees military solutions to most problems, they are merely seeing what comes naturally, through the windows of their minds' eyes. When some college co-eds complained to the editor of *Cosmopolitan* about the magazine's covers that "real women don't look that way," the editor innocently responded, "Oh, but they *can*, if they just follow our beauty tips!"

Our view of the world outside is often seen through a veil of images of ourselves.

One should not be surprised, therefore, when the Senate refuses to do anything to trim down Social Security, but decimates immunization programs and school lunch budgets as easily as taking candy from babies. Most of the powerful senators are old. I am not surprised that the Reagan administration supports policies that benefit the rich—not when some ten Cabinet members, by one count, are millionaires.

What is surprising is the convincing reality of even the most obvious reflections. Bob Miller has built a piece called *Floating Symmetry*, but more commonly known as the Anti-Gravity Mirror. It is nothing but an ordinary, everyday, large, free-standing mirror, like one you might find attached to a closet door in your own house. If you walk up to the mirror and press against it so that the edge of the mirror vertically dissects you through your belly button and your nose, only one half of your body is visible. But you *appear* quite whole to an observer, for the visible half is reflected in the mirror, making two identical halves. If you raise your visible leg, its reflection also rises in the mirror, and you appear to float effortlessly in midair. Even though it's clearly an illusion, people gasp and break into uncontrollable laughter when they see it. They react, in other words, just as they would if the image were real.

Floating Symmetry.

We make remarkably little distinction between objects and images. The real trick to the Anti-Gravity Mirror is that there's no trick except the same "illusion" you see every time you look in a mirror. No one is fooling us but ourselves. Reflections are particularly powerful in this respect because they move images around without changing their shape, so it's hard to see that anything is amiss. Magicians use this to great advantage in making heads float on silver platters, and rabbits leap out of empty hats.

Reflections can move images around without changing their shape.

All science, to a certain extent, is mirrorlike in that we use human perceptions and prejudices to judge the shape and reality of both human and nonhuman things. We are always imposing our view of nature on the nature we "objectively" study. But if all knowledge is circular, it's more like a spiral—a circle that moves along. There seem to be enough cross-connections and concrete pieces of evidence to make some of what we know about the universe "real" (if vastly incomplete). Still, much of reality is always —by definition—reflective. Or as physicist Adolph Baker wrote in his intriguing book *Modern Physics and Anti-Physics*, "We walk around the universe, and, like Winnie the Pooh and Piglet, whenever we encounter our own footprints we say, 'Aha! Someone has been here.'"

Focusing behind *a reflecting surface can open up a new world of images.*

Reflections are not always obvious, of course. But you can almost always find them if you know where to look. Examine, for example, a picture on the wall that is covered with glass or plastic. If you focus on the surface of the picture, you see the picture. But if you focus *behind* the surface, you may be faced with an image of yourself. All reflections (as Alice knew) are behind the looking glass, not on its surface. In fact, a flat mirror image is exactly as far behind the surface as the object is in front of it. This is dramatically proved in Bob Miller's *Everyone Is You and Me*. When two people sit at equal distances on opposite sides of a window-mirror, the *reflected* face of the person sitting in front of the mirror occupies the same space as the *actual* face of the person sitting behind it. (Window-mirrors, by the way, are everywhere. Most windows given the proper illumination turn into mirrors, so that you can easily combine your image with that of a department store mannequin, or see the flames from your fireplace appear to float several feet above the street.) To see the reflected image, however, the reflecting surface must be both clean and clear. If it's obscured with dirt or other extraneous information (like a picture), we tend to focus on the surface and miss the deeper reflection altogether. Like so many other things, reflections are always in our field of vision, but much more rarely in our line of sight.

Reflections are remarkably omnipresent, however. Almost everything reflects something from the proper angle. An echo is the image of your voice as it reflects from a canyon wall. Water waves bounce off the shoreline, reflecting its contours back into the ocean. You can make a mirror out of a book or desk top by peering over its edges at a very shallow angle—just as children learn to make mirrors out of ponds when they skip stones. Guy Murchie tells of a lake in the Alps so smooth that marksmen aiming their bullets at the water are able to hit their targets on land, just as almost anyone can learn to skip stones off the surface of a quiet lake or pond.

A crinkled mirror scatters light like the many small surfaces in snow.

Just about everything we see or hear, for that matter, is reflected from something else. Not all these reflections are mirror-like (like echoes). But anything that doesn't glow with its own light must bask in the light of something that does. When you turn on a lamp in a dark room, the light bounces around from the walls to the furniture and finally to your eyes. Everything you see in the room is really reflected lamplight. Even the red of the couch is a reflection of the red hidden in the spectrum of the white light from the bulb; when the white light hits the red fabric, all the colors except red are absorbed and the leftover red is reflected to your eyes. The rough white walls of the room consist of a kaleidoscope of tiny surfaces—like snow—that multiply lamplight into bright white. If you could pick out just one of these surfaces, you could see that it reflects an image just like an ordinary mirror. But the confusion of so many images melts into white. In the same way, a smooth piece of aluminum foil acts like a mirror until you crinkle it up; the more you crinkle it, the more it reflects like a white wall.

Water waves act like tiny parallel mirrors, smearing reflections of lights into long streaks.

This effect is most familiar on the slightly wrinkled surface of water, where the reflection of the moon or a streetlight is stretched into a long bright streak by the multiple reflections of parallel waves like so many lined-up mirrors.

During the day, our universal "lamp" is the sun. It spreads

light as it reflects from treetops and clouds and bits of air. If there weren't any air for the light to reflect from, the sky would be black. Astronauts travel in the dark. Our night light is the moon, which reflects the light of the sun and also reflects the earth's own light back to us: this explains why you can often see "the old moon in the arms of the new"—the dimly glowing image of the former full moon nestled in the thin sliver of the next. Reflections make the phases of the moon and, indeed, make most things visible. The sunbeams streaming through trees and bathroom windows remain unseen until early morning dew or a steamy shower fills the air with reflective droplets that turn them into visible sunstreaks.

If the air in The Exploratorium weren't filled with reflective dust particles, these sunbeams would be invisible.

Reflection even made the atomic nucleus visible to Ernest Rutherford.* It is still a major tool in unraveling the innermost secrets of the atom. When Rutherford first bombarded a sheet of thin gold foil with subatomic particles and saw that some of them bounced almost straight back, he said he was as surprised as if a cannonball had been reflected from a piece of tissue! Today, people routinely use reflections of radio waves and sound waves—radar and sonar—the better to see all kinds of things.

Polaroid sunglasses filter out light that vibrates horizontally, but they let through light that vibrates vertically. Two Polaroid filters at right angles to each other block all the light.

Curiously, reflections sometimes eliminate as much as half the information in the original light, although we rarely notice it. For reflections can polarize light. That is, they take light that vibrates in many directions—horizontally, vertically, and in between—and reflect only those rays that vibrate, say, vertically. Polaroid sunglasses eliminate glare because glare is bright light that has been reflected from horizontal surfaces such as streets and water. The

*See "Seeing Things."

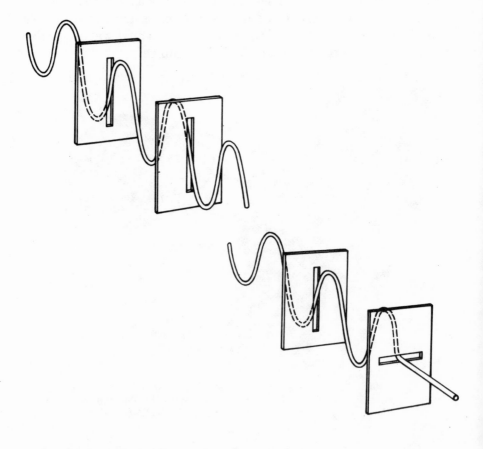

The rope on the left is polarized to vibrate only vertically. The vertical vibration can't pass through the horizontally placed "Polaroid filter" on the right. Many reflections are polarized, which is why Polaroid filters can selectively cut them out.

filters in the glasses cut out only the horizontal vibrations. If you turn your head on its side while wearing Polaroid glasses, you'll notice that horizontal surfaces (like streets) light up again, while reflections from vertical surfaces (like windows) get darker.* Bees perceive the polarization of scattered sunlight, and use it to get around. But to humans, the direction of polarization is beyond our perceptual abilities. We would certainly notice it if someone painted all our windows red and thus eliminated all the green light. But a reflection can subtract as much as half the light and leave us none the wiser.

One of the most enchanting things about reflections is their inherent symmetry—the fact that the image is symmetrical with the object. Symmetry is a powerful and beautiful balance that pervades nature. For every left, there is a right; for every yin, a yang; for every color, a complement; for every particle, an antiparticle. Yet symmetry to a scientist is not exactly the same thing as symmetry is to other people. To most people, a snowflake, for example, is highly symmetrical. But to a physicist, a billiard ball is the ultimate in symmetry. The higher the degree of symmetry, the more you can turn something through any angle and still have it look the same. Mirror images make pleasing symmetries because you can't tell the reflected image from the original object.

People are also highly symmetrical in many respects. We have right and left hands, and feet, and right and left gloves and shoes to go with them. Yet we have hearts only on the left, and appendixes only on the right. For that matter, it turns out that every living thing from elephants to algae is composed of left-handed (or left-spiraling) protein molecules. Technicians can make right-handed sugar in the laboratory, but of course it's not fattening, because it can't be digested by us left-handed beings (or beings constructed of left-handed chemistry, that is). When protein molecules of *any* kind are manufactured in the laboratory, they randomly distribute themselves between the left-spiraling and right-spiraling types. The fact that all naturally occurring proteins spiral to the left probably means that we are all, algae and elephants, descended from the same left-spiraling protein.

Things aren't always symmetrical in the nonliving, or physical,

*This does not apply to metal surfaces, like mirrors.

world, either. Until very recently it had always been generally
assumed that everything that happened in the physical universe
was highly symmetrical; that physics didn't discriminate between
left and right. Then in 1957 Mme. Chien-shiung Wu made the Nobel Prize-winning discovery that there was a left/right difference
in the way things decay radioactively. And the question of symmetry in the physical universe is hardly answered—and in fact, in
many respects the questioning has only begun.

As for the social universe, symmetry offers as much food for
philosophical thought as the related phenomenon of reflection.
"Symmetries," Baker writes, "are not the exclusive province of
artists and scientists. The very process of juxtaposing and examin-
ing real or supposed differences between individuals, groups, par-
ties, races, nations is itself a search for symmetry or lack of
symmetry." To prove this point, he offers a series of quiz questions
at the end of his chapter on symmetry. The question "If a mirror
turns right into left, why does it not also turn up into down?" is
followed by the question "How do black nationalists differ from
white supremacists?" And so on. Posing things in terms of symmetrical situations is often the only way to get people to see the flip
side of their own perspectives. When a man insists on calling
women girls, the best way to cure him is to refer to him as a boy.
When the doctor calls you Sally, call him John. Symmetry really
means a kind of sameness, an absence of distinctions. Einstein's
theory of relativity doesn't mean that "everything is relative" as
much as it means that "everything is symmetrical," which is a very
different thing.*

The amount of symmetry you see in a situation, of course, can
be highly subjective. If you are color-blind, the red light looks the
same light as the green light and you can't tell the difference between the command to stop and permission to go. You can't tell
which way a boat is moving at night because you can't distinguish
the port (red) lights from the starboard (green).

My artist friend, Bob Miller, is primarily known for the wonderful things he does with shadows. He is often asked to take people on what has become known as Bob's Light Walk; it starts

*See "Relatively Speaking."

The circular spots of light that seep through the irregular spaces between the leaves of a tree are images of the sun. Once you start looking, you'll see them everywhere.

outside, in the sun, under the shade (the shadow) of a tree. Bob lifts his hands high above his head and crosses the fingers of the two hands, so that the spaces in between them make a network of irregular holes. But there on the ground, the bright spots of light that get through the holes are perfectly round. They are shimmering images of the sun. The hole between his fingers serves the same purpose as the hole in a pinhole camera. These are the same sun images you see when sunlight filters through the irregular spaces between the leaves of a tree and scatters on the ground like so many spilled gold coins.

Many people don't believe at first that these circles of light are actually images of the sun. (I must admit I didn't.) The only way to

convince yourself is to go out into the sunlight and lift a hole
(punched in cardboard, or fashioned from the fingers of your hand)

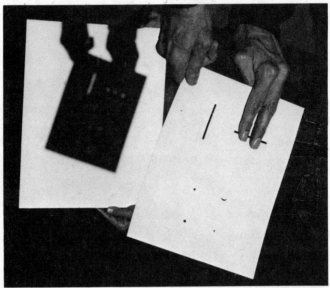

*No matter what shape hole you use (as long as you are far
enough away), the images are round like the sun.*

high above your head and watch the sun shine through it. The hole
can be square, triangular—any shape you like. And still the image
on the ground will be perfectly round like the sun. If the lighting is
right, you may even see a sunspot on the surface. And if there
happens to be a solar eclipse, the round images will slowly transform into crescents. Bob likes to point out that what we call empty
space is filled with images of the sun, and also of everything around
us. If it weren't, we wouldn't be able to catch the image with the
hole that is our pupil when we "see" something. A pupil is a "pinhole" that serves much the same purpose as the holes in Bob's
hand; it lets through a little bit of light while blocking out the excess that would simply blur everything. But if Bob takes his hand
away—or you take your pupil away—the "image" is in a very real
sense "still there," waiting to be plucked out of the air.

If an image is a selected strand of information-bearing light
rays, then what is a shadow? A shadow is a place where a similar
strand of rays is blocked. Often Bob finds a fuzzy shadow on the

ground underneath the tree and plucks a single sun image out of the air above it by making a single hole in his hand. And there on the pavement you can see a clear sharp image of the sun with a dark tree branch crossing in front of it. But something truly remarkable happens when he holds up a small dark spot to cast a shadow, instead of a hole to catch an image. There on the ground is a single, round, dark shadow of the sun—and there crossing in front of the dark sun is a bright, white image of the tree branch.

All the information contained in a scene fits into the tiny hole that is the pupil of your eye; if you take your pupil away, the information is still there.

A shadow, in other words, is a missing image. But it is also a complementary image in the same way that a complementary color is what is left when you subtract one color from white—in the same way that night is the complement of day. It gives the same information as an image, only in a complementary form. In one of Bob's exhibits, a pinhole shadow of a yellow sun, red house, and blue cloud shows up as a purple sun, green house, and orange cloud. Through a whole series of other sculptures and exhibits, Bob has shown that shadows can contain all the information in the light that casts the shadow in the first place. The shadow is every bit as rich in images as the light itself.

Thinking about shadows this way has made me think twice about some of the missing images that inevitably seem to mar people's lives. If you ever feel a slight twinge of regret that you didn't go to law school or continue your ballet lessons, perhaps it will help to realize that the things you leave out define you every bit as much as the things you include. For me, merely having two children and a full-time job is enough to make me a shadow of my former self. But life in my own shadow is hardly an empty place.

Shadows are the contrasts in our lives we need to throw things into relief. Without shadows, a ball looks like a flat, circular disk; a face flattens into a shapeless cartoon; it becomes hard to tell up from down or in from out. A common optical illusion involves taking, say, a photograph of the craters on the moon and turning it upside down, so that the apparent lighting changes and the shadows now appear on the "wrong" side. Suddenly the craters turn inside out, becoming mountains. Shadows give things shape.

I asked some friends if they could think of things in their lives that they didn't do or don't do that helped to define who they are— much like shadows. One, an artist, said he decided not to go to medical school. Another said, "I decided not to get a nose job." And as I looked at her, I could see that in truth she would have been an entirely different person if she had altered the character of her striking face. Still another friend said that she had decided not to learn how to fix her car, even though it seemed to be an important (not to mention fashionable) thing for women to do. "I decided I wasn't interested in cars," she said, "and I'd rather concentrate on things I like and do well and leave the cars to the mechanics." It reminded me of an editor who used to say (every time he felt the necessity to paint, or undertake plumbing chores around the house), "Writers write, editors edit, and painters paint." What he meant was that he wanted to use his energies in those areas in which he was both interested and able.

Shadows are still, well, shadowy. Somehow they have gotten stuck with a negative connotation, a bad name. Yet we live in

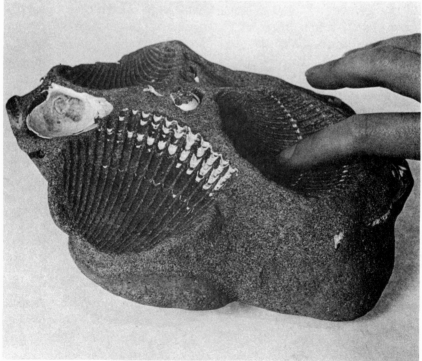

*Shadows determine the shape of things; the indentations turn
into mounds when you change the lighting.*

shadow half the time. Night is the shadow of day—the earth's own shadow cast on itself. If shadows are negatives, then they are certainly not a "lack." After all, it is "negative" electricity that runs through wires and powers our homes and offices. This negative electricity is contained in a very real particle called an electron. It is no more or less real or less "positive" than its antimatter equivalent—a positive electron known as a positron. Indeed, what we call antimatter might be the everyday matter of some unknown universe. The only reason we call it anti- is because in one critical sense it is opposite from the matter that makes up ourselves. When particles are produced in accelerators out of bursts of energy, they always come in pairs—as equal numbers of particles, and "anti" particles.* Making our peace with negatives seems especially important these days when even the empty vacuum is said to be teeming with energy and even "holes in nothing," and when the omnipresent computer stores as much information in the "zeros" of its digital code as in the "ones." Nothing, it turns out, is very much of something. Or as the Greek Leucippus said: *"What is is no more real than what is not."*

The loss of a shadow (as Peter Pan knew) is a loss of an important part of ourselves. Light without shadows contains so much information that it contains no information. If you didn't have an iris to shadow all but a narrow strand of light, you would see nothing but blinding white. (People without eye pigments are often virtually blind.) You can't play a tune on the piano if you hit all the keys at once, and you can't see the sun images on the ground unless you cast a shadow with your hand or stand under the shadow of a leafy tree. Even movies are essentially shadows. The film that passes in front of the projector *subtracts* images from the white light. The film itself does not add information in the form of light, it subtracts it in the form of shadows.

Even clear things, like lenses and prisms, cast shadows. If you hold up a lens or a pair of eyeglasses or a prism to a point of light (any nondiffuse light source will do), you can clearly see that it transports a bright image of the object from one place to another, leaving a shadow in its wake. When a prism spreads out white light

---

*This is not true for some kinds of force particles, which can be created singly. See "Forces, Motives, and Inertia."

into colors, it is really casting shadows. Each color appears in a place where the other colors are not; each color lies in the shadow of all the rest.

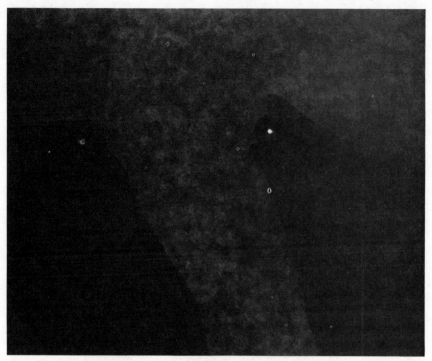

A clear glass marble bends light toward the bright spot in the center, leaving a dark shadow on the outside.

And light isn't the only thing that can be blocked to cast shadows. An umbrella or a building casts wind shadows and rain shadows as well as light shadows. A glass skylight casts a rain shadow while letting the visible sunlight through.

Few shadows block out *everything*. Polaroid sunglasses shadow only the horizontally vibrating light; they cut the glare, but actually improve your vision. Shadows are really like filters. Rather than obscuring things, they select things—just as the coffee filter "shadows" the coffee from the grounds. When you pour the cooked spaghetti and water through the colander, the colander creates a shadow for the spaghetti but not for the water—and a tea strainer casts a shadow only for tea leaves. Shadows let you filter the information from the noise, the important things from the distractions. Without them, you wouldn't be able to see the image for the glare. In fact, the screening or inhibition of extraneous nerve signals is as important to perception as the sending of those signals in the first place.* Shadows are shades that protect us from too much of a good—or bad—thing. People who don't filter anything out of their lives often wind up murky, like unfiltered coffee. A life that has everything remains unrefined.

Shadows are created, of course, by obstacles. And often, the shape of the shadow can tell you a lot about the obstacle that cast it. Aristotle looked at the round shadow of the earth on the moon during an eclipse and deduced that the world was a sphere. Film is the "obstacle" in front of the projector that creates the picture, just as your bones are the "obstacle" in the path of the X rays that create the X-ray pictures. As the obstacle gets close to the light, it blocks out more and more and the shadow grows large and scary,

The dark shadow in the center is the "umbra," where all the light is blocked. The softer surrounding "penumbras" are places where no light gets through.

*See "Seeing Things."

but ill-defined. As the obstacle moves away from the light toward a wall or screen, the shadow gets smaller but sharper.

But the form of the shadow depends on more than the obstacle; it also is shaped by the light (or other information) itself. If there is only one light in someone's life (or in any situation) and something or someone blocks it out, it leaves a large and dark shadow. But if there are other lights, then they can fill in where the shadow leaves off. The single, sharp, dark shadows created by a single small light source and/or an obstacle up against a wall are called "umbras." The softer surrounding shadows created by more diffuse, spread-out light are "penumbras." If you look at the umbras and penumbras of shadows, you can sometimes learn much about the numbers and shapes of the lights illuminating a particular space.

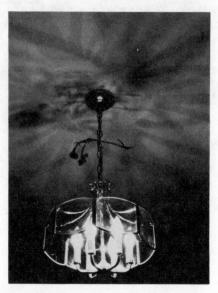

Multiple lights make multiple penumbras.

Finally, shadows reveal many things about the surfaces they fall on. Shadows that fall on curved surfaces are curved, and shadows that fall on curbs or stairs are zigzagged. The spherical shape of the earth was also deduced from the fact that shadows falling on it at different places and times have different lengths and shapes.

It is impossible, however, to cast a shadow on something that glows from within. You can't cast a shadow on a fire, or a star, or a red-hot poker. Shadows can be cast only on surfaces that bask in the reflected light of something else. (Perhaps this is why teachers and parents are always telling children that when honor and respect and happiness come from within, nobody can take them away from you. Of course, somebody brighter—or meaner—*can* take them away from you. And you can even cast a shadow on hot coals if you first illuminate them with a brighter light still.)

The shape of a shadow depends on the surface it falls on.

Shadows are full of information. But in the end, they are only projections. True, a shadow in space fills a three-dimensional volume in the same way that an image in space is a solid (if ethereal) form. The shadow we live in at night isn't a flat shadow on a screen, but a huge cone of darkness that spreads far into space.

Without these shadow spaces, we would have no place to hide, to move about unseen, to enjoy a private life.

Shadows are three-dimensional; you can hide in them.

But on a surface where we normally see it, a shadow is a two-dimensional image just like the two-dimensional image we "see" inside our eyes. Shadows lack depth, and so can be easily misinterpreted. The shadow of a cylinder can look like a circle or a rectangle or an oval, depending on how you play the light. This aspect of shadows allows us to use them to create wonderful shadow creatures out of the fingers of our hands. But like other two-dimensional images, shadows tell only one point of view.

12.
SYMPATHETIC
VIBRATIONS

Metaphor is truly a marvelous thing. People speak of being in tune with the times or out of tune with each other. They speak of sympathetic vibrations, and of being on the same wavelength. They speak of ideas that resonate and descriptions or anecdotes that ring true. I wonder how many of them know that all the while they are talking physics.

Resonance is the physics lesson all children learn the first time they try to pump themselves on playground swings. The trick, they soon learn, is timing. Pushing forward or leaning backward at the wrong place or time gets them nowhere. The thrust of the pump has to coincide with the natural rhythm of the swing. The key to resonance is pushing or pulling in time with the way the swing (or anything else) naturally wants to go. It is the synchrony of many small periodic pushes that work in unison to add up to a much larger one.

Until recently, I had always lumped resonance with other broad if rather boring phenomena (like sine waves) that only a physicist could love. Then one day I was reading an account of the exciting discovery of the so-called J/psi* particle in the Stanford Linear Accelerator Center's newsletter, *Beam Line*. (It was exciting because it was the first evidence of the property physicists whimsically have taken to calling "charm." It was also exciting because the particle popped up simultaneously at two competing laboratories—Stanford, where it was named psi, and Long Island's Brookhaven National Laboratory, where it was called J. In a bit of lingering institutional chauvinism, some Stanford scientists still insist on calling it the psi/J.)

In any event, the thing that caught my eye was that almost nowhere in this account did the writer refer to the new subatomic species as a particle. Usually, it was hailed as an unusually sharp "resonance."

Resonance?

It is no surprise to a physicist that a *thing* can be a "resonant state," but it was certainly news to me. Little did I know that resonance was behind the very nature of matter—not to mention the sound of music, the color of autumn leaves, the rings of Saturn, the spectral lines that write the signatures of stars, and according to one source perhaps even the evolution of life. It accounts for

*The J/psi is one kind of particle that may emerge briefly in the "debris" of highly energetic collisions between particles and antiparticles. It does not itself exhibit the quality (actually, quantity) physicists call "charm," but it is made up of a combination of charmed and anticharmed particles that cancel each other out. J/psi particles are not the stuff of everyday earthly things. Rather, they are evidence of the kind of matter that exists at very high temperatures, like those that fired the very early universe.

everything from the low whistle of the wind over the Grand Canyon to the sound of a lover's voice. Surely such a pervasive phenomenon was worth looking into. And not surprisingly, it turned out to have interesting (if not necessarily scientific) social and psychological parallels.

Literally, resonance means to resound, or sound again: to echo. Putty doesn't resonate bcause it is too full of internal friction to vibrate; a dropped handkerchief simply falls. In order for something to resonate, it needs a force to pull it back to its starting place and enough energy to keep it going. The trick is to have it resound again and again in a kind of continuing echo, but this requires both keeping friction at a minimum and putting energy in faster than friction can take it out. Two can play this game much better than one because one can feed energy to the other. That's what sympathetic vibrations are all about. When people talk about resonance, they are usually referring to the confluence of more than one action: two partners in a business feeding each other ideas and energy at just the right times and places to get big results. Or the escapement of a watch that lets the spring go just enough and at just the right time to give the pendulum (or crystal) the push it needs to keep it going.

Putty and handkerchiefs aside, the universe as a whole is a remarkably springy place. Planets and atoms and almost everything else in between vibrate at one or more natural frequencies. When something else nudges them periodically at one of those frequencies, resonance results. Soldiers marching in step with the natural frequency of a bridge can cause it to collapse, which is why soldiers break step when crossing bridges. The Tacoma Narrows Bridge in Washington State was toppled in 1940 by a resonance caused by wind. The ill-fated prop-jet Electra developed an unfortunate habit of falling apart when the rhythm of the rotating propellers matched the natural frequency of the wings. Nikola Tesla, the eccentric inventor of AC current, became so obsessed with a phenomenon called electrical resonance, he thought he could use it to split the world. In fact, recent studies suggest that giant icebergs are splintered by the resonant force of gently lapping ocean waves. Resonance is even responsible for the forty-foot tides in the Bay of Fundy. My friend the physicist insists that through a similar series of well-timed pushes on water, any determined child (or adult) can empty a full bathtub in a single cycle, once the "tide" has been built up high enough.

The power of resonance comes literally from being in the right place at the right time. For it to work, there has to be a harmony between what you're doing and the way something (or someone) wants to go. The almost eerie purity of laser light comes from the fact that all the atoms in the excited gas are poised just so that a gentle nudge of energy will cause them to give off light in patterns exactly aligned with each other. In much the same way, the gravitational pulls of Saturn's four inner moons happen to be in harmony with the natural period of rotation of some particles in the rings at certain distances from the center. This period at one point coincides exactly with one third of one moon's period, one half of another moon's period, one quarter of still another's, and so on. The combined force is enough to push (or pull) all particles out of those places, creating the gaps in the rings. In fact, if it weren't for resonance, they wouldn't be rings at all, but a single solid disk.

Resonance, in other words, allows a lot of little pushes in the right place to add up to big results. Particle accelerators use this principle of "a well-timed kick in the pants," as one physicist put it, to nudge electrons and protons almost to the speed of light. But it is also a familiar phenomenon in everyday life. A lot of little pushes in an already angry crowd can lead to a full-scale riot. A lot of little digs over the dinner table can lead to divorce. A few otherwise insignificant failures at a low point in someone's affairs can lead to depression or even suicide. Sometimes clever politicians exploit this property to play on the public's natural dissatisfaction with inflation, or taxes, or spending—and get a large resonant response with a very small input of energy or new ideas. But with no additional input of energy, the response tends to dissipate rather quickly.

Even hula hoops, I recently rediscovered, are a good example of resonance. In the first place, the only way to keep the hoop revolving is to push it in time with the natural frequency of the rotating ring—like pumping a swing. But in addition, whoever invented hula hoops happened to hit the public's pocketbooks and imaginations at just the right time. Had hula hoops come along, say, during the 1960s or 1930s, it's unlikely that they would have caught the crest of such a giant wave of popularity. (See "Waves and Splashes.")

But resonance is far more than a brutal amplifier. It is also music to our ears. The opera singer's aria is an ode to resonance; each pure tone that fills the opera house is the tiny vibration of a vocal cord amplified by the shape of her chest and throat. (The same pure tone, of course, can shatter a crystal glass.) A violin bow slips along the string, catching it imperceptibly at precise intervals that push it at the proper time to keep it vibrating. The body of the instrument vibrates in tune to a rich range of harmonics. To play a flute, you set the air inside resonating at many different frequencies, depending on how far the sound waves travel between the mouthpiece and the fingerholes. The holes are placed to pick out those tones that correspond to the standard musical scale—but how you blow determines whether your flute will resonate with the notes that are purest and loveliest.

In this sense, resonance may provide a good analogy for the process of learning. It often takes a lot of little pieces of information to add up to a deep understanding, and a lot of little insights to add up to a great idea. But these little bits and pieces need to come at the right time and strike you in the right way. If the information comes along before you are prepared to offer feedback, the input simply dissipates like a swing pumped the wrong way. Adding energy without timing gets you nowhere. If you pump at the wrong time on a swing, you are opposing the natural pattern—bucking the current, so to speak. You sap energy instead of adding it. The flute player who sets the air vibrating at a frequency not natural to the flute doesn't get a note at all, but so many different frequencies that the result is an amorphous, irritating hiss.

Being out of tune is always irritating because the energy you put into your efforts doesn't get you anywhere. It goes against the grain instead of with it. It steals your harmony and leaves you noise. You can be in or out of tune with yourself—like the flute and the violin—or in or out of tune with others. Two flutes played beautifully but slightly out of tune with each other can be much more unpleasant than one played badly. Partners in a marriage who are perfectly in tune with their professional or personal lives can be badly out of tune with each other, too. Members of different generations often find it difficult to stay in tune with both their peers and the other members of their families, and it is rare to find yourself on the same wavelength with the same people at home, at work, and on vacation.

All these examples of resonance have the same essential property, but they achieve it in at least two different ways. In one case, you have two things pushing in time with each other—like a person who pumps in time with the natural frequency of a swing, or the gentle pushes of air set in motion by one tuning fork that can nudge an identical fork nearby to vibrate in tune.

In another case, however, you can get the tuning fork vibrating by whacking it, say, against your shoe. This kind of resonance feeds itself. Like the strike of a bell or the wind on the Tacoma Narrows bridge or the tension in the spring of a clock, the energy to feed the vibration comes in a single push—sudden, or steady. But the shape of the resonant object itself is such that it doles out the energy at exactly the right time: the bridge twists just in time to catch the next resonant gust of wind; the escapement of the clock lets the energy wound up in the spring "escape" in time with the natural frequency of the pendulum or crystal. Either way, however, a so-called resonant system can hold on to its energy for a relatively long time. This makes it easy to pick out the resonant systems from the nonresonant ones.

Say you walked along a path strewn with pebbles and bells, and kicked them out of your way at random. The bells would ring while the pebbles wouldn't. Why? Because the pebbles would reflect the energy of your kick every which way as they flew off in different directions. But the bells would be able to feed the energy back into themselves by virtue of their natural springiness. The bells would be able to contain the energy long enough to "ring."

In fact, the most important property of resonance is this ability to act as a precision tool that can separate the bells from the pebbles, pluck one responsive vibration from a sea of others, a single tune out of a confusion of white noise. Resonance gives things character: the difference between the music of a violin and a flute, the difference between the voices of a man and a woman, the difference between the clatter of a tennis racket and the plop of the ball, are subtly shaped by sympathetic vibrations. When disturbed, each object resounds and vibrates only to its natural set of frequencies, which together form its special sound. All other vibrations are canceled, or sent flying off in random directions. There is no mistaking the voice of your child because the rush of air from the lungs that starts out as just so much noise is selectively amplified by a particular configuration of mouth, nose, chest, and throat

Each pipe resonates to a specific frequency and picks out that one sound from the wide spectrum that fills The Exploratorium. Anything that resonates responds to one frequency above all others.

to sound in a particular way—just as the tuner on your radio amplifies only one narrow range of frequencies at the expense of all others. The others are simply scattered.

The character of all these things, in a sense, is determined by what they *respond* to. And so it is with people: the things that shape your character—the books you read, the friends you spend your time with, the occupation you choose as a career—all are things that respond to resonant notes in you. In the same way, plucking a piano string will cause a series of other strings to start vibrating—but only those in perfect harmony with the string you originally plucked. By looking at the responding strings, you can deduce a great deal about the note that struck the resonant chord in the first place. And if you strike a different note, a different set of strings will respond.

Plucking a single piano string sets a whole series of other resonant strings in motion.

This very same property, in fact, colors everything you see. Sodium lights are yellow because sodium atoms vibrate with those frequencies your brain perceives as yellow. Mercury atoms vibrate with a bluish light; neon atoms send out vibrations that reach your brain as "red." Yet if you look at sodium or mercury or neon through a prism, you can see that the colors are not the result of a single "note," but rather of the atom's characteristic "chord." The sodium spectrum contains two sharp yellow lines—frequencies of light emitted by specific vibrational states of the atom.* But the sodium spectrum also contains a thin line of red and of green, and neon red also contains some green and yellow. The sum total of all the lines gives you a clear picture of the possible vibrational states of a particular atom, and so it tells you exactly what that atom is. A physicist can see that a particular set of blue, yellow, and green lines spaced in a particular way comes from mercury atoms, just as a musician can hear that a particular set of harmonics means a note is coming from a violin.

These telltale harmonics of atoms are clues to the kinds of elements that exist in unreachable realms, like the surfaces of stars. Sometimes the elements show their colors by the frequencies they emit; in other cases, they give themselves away by the colors they absorb. But since atoms both absorb and emit the same "chord" of frequencies, the only difference is whether the telltale fingerprints of the atom show up as a spectrum of colored lines, or as thin dark lines or shadows in an otherwise continuous spectrum.

Guy Murchie, in *Music of the Spheres*, explains the presence of these shadows with an appropriately musical example:

> If you have three tuning forks in a room, each keyed to a different note in the major triad C-E-G, and this same chord is played in the next room with a large peephole opening between the two rooms, your three forks will start to hum the chord by sympathetic vibration through the air. But if another E fork is placed exactly in the peephole and the experiment repeated, the original E note will be found missing from the resultant dyad, which will now consist only of C and G, the middle fork being silent because the new E fork in the peephole hums in its stead, having absorbed most of the energy of E frequency as it tried to pass through the opening.

White light from a star that passes through surface gases containing an element that strongly absorbs green will arrive on earth with a sharp shadow in the green part of the star's spectrum. This

*See "Quantum Leaps."

kind of long-distance chemical analysis has revealed that we and
the stars are fashioned of the same stuff.

Down here on earth, resonant absorption colors everything
from sports cars to begonias. The pigment molecules in the skin of
a McIntosh apple absorb the parts of sunlight that vibrate harmonically in the frequencies we see as blue and green; the rest of the
light is reflected, and we see red. Chlorophyll molecules in green
leaves vibrate to the tune of red and blue and absorb them, reflecting the leftover green; the same leaves absorb green and reflect
autumn colors in the fall. Ultraviolet light vibrates harmonically
with the molecules in glass; the visible light gets through, but you
can't get a suntan unless you open the window. The ozone layer in
the atmosphere, like suntan lotion, also absorbs much of the resonant ultraviolet vibrations from the sun and protects us from potentially damaging light.

The red in the apple, that is, is reflected to your eyes because
red is *not* resonant with the red pigment molecules in the skin; the
red light is reflected back to you like the pebbles. The resonant
vibrations of blue and green, on the other hand, are absorbed by
the skin. Their energy "rings" inside the skin for a while, and eventually is reradiated in all directions as heat. In the same way, light
that vibrates sympathetically with elements on the surface of a
star "rings" for a while, or is absorbed and reradiated, before it can
get to you. The places where the light is absorbed are the dark
spaces you can see in the star's spectrum.

Even a rainbow is a resonance phenomenon. The colors of
white light passing through a prism or a raindrop spread out because the colors near the violet end of the spectrum are more
nearly resonant with the glass or water molecules than the colors
at the red end. The closer a color is to perfect resonance with the
glass molecules—that is, the more in tune the two vibrations are—
the longer it lasts. The purer the resonance, the longer—like a
good bell—it rings. Violet light "rings" longest, and therefore
bends most, as it passes through a prism.

Resonance, in other words, determines what we see, and
what's reflected; what goes right through, what gets stuck, and
what sinks in. It is the difference between visible and invisible;
between transparent and opaque. Metals are opaque because their

many freely moving electrons can vibrate to just about any frequency—and so absorb them. The ability of these same free electrons to reradiate all those frequencies explains why metals make good mirrors. On the other hand, almost everything is transparent to a radio wave because almost nothing resonates in radio frequencies. You can hear your radio (and TV) signals right through the thickest walls. One of the few things that can resonate in radio frequencies is the free electrons in metal, which is why radio aerials and antennas are made of metal. It also explains why you lose radio reception while driving over a heavily girdered metal bridge. A single antenna is exactly the right length (usually one-quarter wavelength) to send out or receive only one resonant frequency (or a narrow spectrum of frequencies). A metal bridge offers so many paths to free electrons that an incoming radio signal sets them moving in a random way; the signal gets absorbed long before it can reach the antenna of your radio.

Infrared or heat radiation is resonant with almost everything—which is why everything absorbs heat. Glass and paper are invisible to shorter-wavelength microwave radiation, however; they don't "see" the heat in your microwave oven and therefore don't get hot. Radar is a kind of radio radiation that finds clouds semi-invisible but airplanes opaque—like looking through a smoky glass. Some frequencies of radar can see through solid ground to find pipelines ten feet underneath. Lately, satellite telephone transmissions have been running into obstacles because the microwaves that carry them are approaching the size of raindrops. When they vibrate in tune, your voice gets stuck in the storm.

Sometimes resonance turns things into one-way doors, or radiation traps. Glass can be just such a one-way window for light. Visible light from the sun passes through a glass window and is partially absorbed, say, by someone inside wearing a red dress. The red light reflected from the dress can pass right back through the window, but the other colors are absorbed by the dress and eventually are reradiated as the lower-frequency light we call heat. But the heat radiation can't get back through the window. It is trapped inside. The result is the so-called greenhouse effect, which is great for warming up greenhouses, but dangerous for our atmosphere. An excess of hydrocarbons from burning of oil and coal are making the sky into a one-way window that traps the sun's heat, warming up our environment in ways that may prove hazardous to everyone's health.

What's visible and what's invisible clearly depends largely on what (or who) is doing the looking. What's visible is whatever you happen to be tuned in to. Tuning in to a radio or television station merely means putting your receiver on the same wavelength as a transmitter somewhere else. In the process of tuning in to one channel, of course, you tune out all the others—just as sodium is tuned in to yellow, but not to violet or green. Teenagers tune in to teenagers, kids in to kids. It is a well-known fact that pregnant women tend to see the world as suddenly populated by huge numbers of previously unseen pregnant women, just as the parent of a two-year-old begins to see two-year-olds wherever he or she goes. If you're not attuned to racist or sexist discrimination, you may be blind to it even when it's right in front of your eyes. Friends or family members who are tuned in to each other can often transmit and receive subtle signals that are completely invisible to everyone else. When you're tuned in to one thing, you tune out the "extraneous noise" that might interfere with the quiet nuances of the signal.

All people—like all things in the universe—can resonate to many different frequencies. But only rarely do you find yourself exactly in tune with something or someone else. When you do, it's the click of recognition that comes from finding a friend who laughs at your jokes; whose understanding goes without saying. Discovering someone who's a good "resounding" board for your ideas is as rewarding as the sound of a well-tuned flute. There's nothing more frustrating than having the thoughts you throw out land with a thud, instead of springing back with that extra surge of energy provided by an insight from someone else.

It is in the nature of most vibrations, however, that they tend to get in and out of tune with each other. Anything that resonates necessarily oscillates; it swings back and forth. Just because two vibrations start out at the same time and in the same direction doesn't mean that they will stay that way; if they reverse direction for the return swing at different times, they will rapidly get out of tune. Each part of a swing has its phases: ups and downs, starts and stops. If your timing is not right, you can easily find yourself up when your partner is down, just starting when your partner has already stopped. Even flutes get out of tune as they warm up. People or things that make beautiful music together must continually be adjusted, fine-tuned. Resonance is a delicate thing.

Still, it's probably not such a bad idea to change our tunes every once in a while. A pleasant and fiftyish woman once visited The Exploratorium and complained that people of her age couldn't relate to the teenaged guides. Of course, she had a point, but then how many teenagers have been permanently put off museums by guides who are fiftyish women? Some people spend a lifetime looking for friends or spouses or hoping for children who are perfectly in tune with themselves. But how much nicer it is to vibrate to a large number of frequencies; how sad it would be to tune in only to yourself. If everyone did it, we would all go through life permanently plugged in to something like a Sony Walkman, humming only our own private tunes. In truth larger patterns of getting in and out of tune with various things and people are often more rich and lasting than any single tune itself.

In fact, it turns out that resonance usually works better in the presence of a little friction—because the resonance is *broader;* it doesn't have to be *exactly* on the mark. It isn't as pure, but it's much more flexible.

Fortunately, it also turns out that most things vibrate not only to a single note, but to a rich spectrum of sounds called harmonics—multiples of the original frequency. Harmonics are the embellishments that take the tone from a single string and give it the rich overtones of a wide range of sympathetic vibrations. Harmonics sound the difference between a Stradivarius and a fiddle.*

This still hasn't brought us back to the question of how a resonance can be a particle. Of course, one really can't explain particles in terms of submicroscopic bells or tuning forks. But one can draw some interesting (if not exactly accurate) parallels.

For example, resonance can make things seem to appear out of nowhere—like rabbits out of hats. The music broadcast by your local radio transmitter seems to spring out of thin air when you tune your receiver to a sympathetically vibrating frequency. Your smoothly running car can suddenly break out in a bad case of the shakes when the cycle of the unbalanced wheel exactly matches

*See "Waves and Splashes."

the natural rhythm of the springs. At, say, sixty-two miles per hour, the wheel is bouncing up and down and so is the spring. But if the spring is still pushing down a little bit while the wheel is pushing up, the motions cancel each other and you don't notice anything. If, however, at precisely, say, fifty-nine miles per hour both wheel and springs (and therefore body) are going up and down in the same rhythm, feeding energy to each other at every turn, a violent shimmy seems to appear where a moment ago there was no motion at all.

Some people have used this analogy to explain how resonances can produce particles. In the elementary universe of particle physics, every energy is associated with a frequency and vice versa. It is part of the natural complementarity of matter that it has both wave and particle characteristics. Since matter has wave properties, it has frequencies, too. Each particle wave, for that matter, has a specific frequency, and that frequency corresponds to a specific energy. Energy (according to $E = mc^2$) equals mass. So in a very fundamental sense, the way something "vibrates" seems to determine what it is. And when the physicists at, say, the Stanford Linear Accelerator Center tune their beams of electrons and positrons so they collide with a burst of energy that vibrates at exactly 7.5×10^{23} cycles per second, then presto! They have created a particle (or really pairs of particles) in much the same way as you can create a tone by blowing with precisely the right energy over the top of a Coke bottle. The explorers at SLAC don't call what they do particle hunting, but rather "resonance hunting."

All analogies break down as you descend to subatomic depths, of course. But you could also imagine the particle/resonances as the bells on the pebble-strewn path. Most kinds of collisions would result in a lot of kicked pebbles—with energy and motion redistributed in all directions. Every now and then, however, you would hit something and it would "ring" for a longer time, because it would have the special property that it could feed energy to itself. You would know that there was something special about it. You might even call it a particle. It is a rather nice thought that the universe might turn out to be essentially just such a symphony of submicroscopic chiming bells.

13. WAVES AND SPLASHES

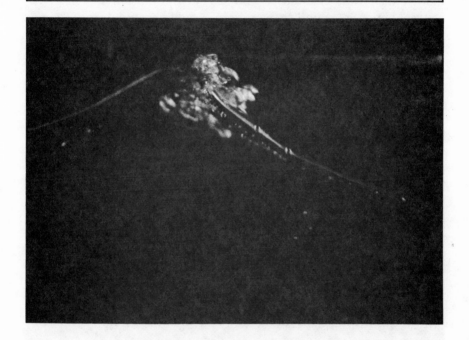

It is said of some people that they make a splash and of others that they make a wave. Language can be so accurate. Logically there is no difference . . . a splash always causes a wave . . . but people who make a splash do not always make a wave. They do so only when the spreading effect of the splash moves other people.
—*Exploratorium* magazine, Winter 1982

A vibration is a wiggle in time; a wave, a wiggle in space. A wave is necessarily a kind of vibration, but you can have a vibration that doesn't make waves. (Something that vibrates in a vacuum tends to play only to itself.) Normally, however, any kind of twang or disturbance in the nature of things spreads out in far-reaching waves of influence. A splash, on the other hand, is a one-shot affair. The Edsel made a splash; Elvis and Einstein made waves. The difference between a wave and a splash is that a wave is a lot larger than itself. It can separate itself from the original disturbance that created it and carry information far from its source, bending around corners, going right through things, sometimes capsizing people or even whole countries in the process. Once out on its own, its strength does not depend on any particular event or leader, but has a power that sometimes exceeds all expectations. It can interact with other waves in ways that make it grow to monstrous proportions—or completely disappear.

Einstein and Elvis made waves.

A wave can do all these things because it is not made of "stuff"; it is a movement of information. A fashion wave can start with a splash, for example. Somebody famous starts to do something different or wear something new and the fad spreads to other people. But once set in motion, the wave moves quite independently of the people caught up in it. The people are only the carriers. The wave itself consists of a pattern of how people pick up the fad and then drop it. As the wave of, say, health food stores spreads from California to the East Coast, the health food stores do not pick up and move. They stay where they are; only the wave spreads. In the same way, waves of thought and feeling spread throughout your body using nerves as electrical conduits. But the nerves themselves stay put. Most waves, in fact, die out rather quickly. Only when there is a continual input of new energy—people picking up a fashion wave, wind fanning ocean waves, fresh sources of electrical energy nudging along nerve signals like so many telephone cable repeaters—do they keep going or even gain strength.

Some splashes make more than one wave. When you drop a stone in water, the air next to the water receives a push, which it passes along to the next bit of air, and so on and so forth like a pulse sent along a Slinky or the wave of motion (and emotion!) that one car nudging another can send through a traffic jam. The push of air arrives at your ears and is heard as sound. The air molecules themselves, of course, move only to vibrate back and forth; it is the push itself that actually travels At the same time, however, the splash starts a wiggle in the water. The up-and-down motion travels through the water, incidentally carrying up and down any sticks, leaves, ducks, or boats that happen to be in its path. The sticks and boats won't travel to the opposite shore any more than do the particles of water. The wave only moves *through* the water, like a rumor through a crowd of people.

What makes waves, in other words, is not traveling stuff but moving signals. Light waves and sound waves carry voices, words, and images. Ocean waves carry information about storms far out to sea, and a tidal wave brings the message that somewhere a crack in the earth has moved. The first domino in a row transmits the fact that it has fallen to the last domino without ever moving itself. What has propagated through the row is a change in the domino's *position*—from vertical to horizontal. What has moved is a *condition*, a state of affairs.

If this material unreality of waves makes them seem nothing more than an abstract pattern, consider that almost everything around you—including yourself—is essentially just such a pattern. What is a person, anyway? Surely not just a collection of atoms and molecules; the ingredients themselves are barely worth the price of the cake. It is the *pattern* of atoms and molecules that both defines a person and makes him or her precious. While the lives of individual cells may vary from a few days (gut and skin) to a lifetime (nerves), the atoms that make up those cells are completely replaced every five or so years. The pattern of matter that makes up your hand is more or less the same pattern (give or take a few wrinkles) that made up your hand five years ago. But the matter itself isn't the same. So is it the same hand, or isn't it? (If sometimes you don't feel quite like your old self, now you know why.)

Even the ultimate pattern that charts the course of all other patterns in a living being—the double helix of DNA—is only, after all, a collection of atoms and molecules. They too can be (and are) continually replaced. Only the pattern remains.*

Life, of course, is dynamic, so we should expect, perhaps, that the patterns would turn out to be more real and lasting than the individual parts. But what about inorganic objects? Surely they are more than abstract patterns. But defining them is not so simple either. "Consider an object," writes Richard Feynman.

> What *is* an object? Philosophers are always saying, "Well, just take a chair, for example." The moment they say that, you know that they do not know what they are talking about any more. What *is* a chair. Well, a chair is a certain thing over there . . . certain?, how certain? The atoms are evaporating from it from time to time—not many atoms, but a few—dirt falls on it and gets dissolved in the paint; so to define a chair precisely, to say exactly which atoms are chair and which atoms are air, or which atoms are dirt, or which atoms are paint that belong to the chair, is impossible.

Eddies and whirlpools and raindrops are patterns of water mol-

*For more on the abstract nature of organisms, see Guy Murchie's *The Seven Mysteries of Life*.

ecules (it does not matter *which* water molecules) that exist almost independently of the water itself. The water flows *through* the forms.* A rainbow is not a thing, but a pattern of light refracted from water droplets—*different* light from *different* droplets for every person who sees it. Each person sees a rainbow unique to a personal perspective; only the consistent pattern of colors in a curved bow deceives people into thinking it's a concrete "thing." The water molecules themselves have properties that depend on the pattern of oxygen and hydrogen atoms that make them up. All the qualities that make water the fountain of life—the stuff of blood, sweat, and tears—come from the arrangement of these atoms, the same arrangement that is reflected in the shape of every snowflake and soap bubble.

A candle flame is a constant flow of different particles; the pattern of particles is more "real" than the sum of its parts.

Even a galaxy of stars is largely an abstraction, in the sense that the individual stars in the spiral arms are continually replaced by new ones. The pattern of stars that makes the spiral even ro-

---

*A beautiful book on this subject is *Patterns in Nature* by Peter S. Stevens.

tates at a different speed than the stars themselves. The sun currently resides between spiral arms, but once it was spinning out on a limb, and (if it lives long enough) may well someday migrate to an inner limb, a home for more middle-aged stars.

Abstractions seem magical because they can exist independent of matter—and also because they can do things that matter itself cannot do. Family traits and traditions can long outlive any individual family member, just as a comet's tail can sweep around the sun faster than the speed of light. It can do this because, as Murchie points out, "a comet's tail does not remain the same tail any more than a stream of water from a hose remains the same water." Any more than a candle flame remains the same flame (it is a constant flow of different particles) or you remain "yourself" for more than a five-year period. In all these cases, the essential reality is the pattern, and not its components.

Murchie asks the question Where were you in 1800 A.D.? Where was the telephone?

> None of us had yet been born. All the elements of our future compositions were in the world but not organized into integral systems. And just as the right combination of thoughts and actions produced the telephone and developed it as a system of communication, so did the right combination of motivations and germ cells produce you and me. Thus an organism of life is basically much the same as an organism of systematic ideas—an abstraction, a new combination, a larger reality.

He points out that this reality obviously "does not need a location—which explains how a lens with a focal length of ten feet can be packed in a box only two feet long."

A candle flame is a flow pattern in space, but many patterns take shape in time—music, for example, or conversation. The patterns of words and notes are far richer than the sum of their parts. All vibrating things—and resonances—are composed of patterns in time, as are patterns of human history. The pattern of evolution is much more concrete than the sometimes fleeting species that make it up—just as a person who continually replaces one friend or job with a "better" one is more deeply wedded to a behavior pattern than to any friend or job in the chain. For the same reason, patterns of discrimination or abuse are often more "real" and therefore easier to document than individual instances.

Historical patterns, of course, frequently travel in waves. Even science itself (or public perceptions thereof) has ridden great waves of popularity only to be plunged in the next decade or century into a trough of disrepute. Reading Daniel J. Kevles's book *The Physicists*, one is almost made dizzy by the force of these waves. Total disdain of physicists as either useless eggheads or helpmates of the devil alternated with a reverence that really seemed to have more to do with fashion: "Physical scientists are the vogue these days," wrote a *Harper's* magazine contributor just after the war. "No dinner party is a success without at least one." Needless to say, the physicists may have had something to do with making the waves in the first place. Or as Ernest Rutherford reportedly said when asked why he seemed to be on the crest of every wave in physics: "Well, I made the wave, didn't I?"*

Perceiving these patterns is so important to human survival that we often see patterns that "aren't there"—in clouds, in cracks in the ceiling, the "man" in the moon. Actually, these patterns are highly subjective. In China, it is a "rabbit" in the moon. The configuration of stars we call the Big Dipper is interpreted as the Plough in England. But one way or another, almost everyone is involved in looking for patterns that add up to a larger reality. Doctors look for patterns of symptoms that spell disease, journalists look for patterns they call social trends, entrepreneurs look for patterns in the marketplace, and scientists look for patterns known as Laws of Nature.

The laws of nature are really observed patterns—relationships between things and events. They have great power because they allow for infinite variety amidst amazing regularity. All people— like all planets and all trees—are cut from the same pattern, yet they exhibit a wide range of individual forms. We are all alike, yet

---

*If the physicists with their atomic bomb were all the rage after World War II, the chemists, with their poison gas and explosives, were the heroes of World War I. My friend the physicist tells of a group of scientists working on the atomic bomb at Los Alamos who speculated (tongue only partially in cheek) that World War III would be the war of the biologists—and World War IV, of the psychologists.

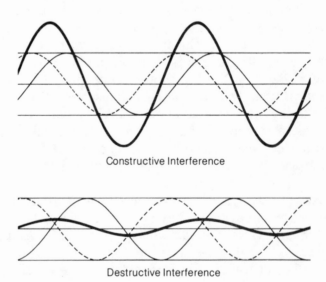

Constructive Interference

Destructive Interference

When sets of two waves interfere constructively, their energy adds up to an area of extra brightness or loudness. Waves that interfere destructively cancel each other; the result is darkness, or silence.

all different. These patterns or "laws" are often expressed as mathematical formulas—something that allows them to reach out and explore far beyond the realms of human experience. A pattern—unlike a person—can safely go to extremes. It can tell you what will happen if you leave your money in the bank for a million years, or what life on earth might have been like a million years ago. It can both extrapolate and interpolate. It can help you to find out where you're going and where you've been. It can tell you what takes place inside an atomic collision, and what matter is like at infinite gravity—as in a black hole.

INTERFERING PATTERNS

Patterns—like people—often interfere with each other. Sometimes the interference is constructive and at other times it is destructive. Two sets of waves interfere constructively (often in resonance) when crest meets crest and trough meets trough so that the effects of the waves add up. When two sets of waves interfere *destructively*, however, one is moving up while the other is moving down; crest meets trough and the two motions cancel each other. The result is nothing. Whenever two somethings add up to nothing, it is a sure sign that you are dealing with waves. Two houses (or even two pebbles) cannot add up to no houses or no pebbles. The fact that two light beams interfere with each other in ways that produce bands of darkness stood as firm evidence for centuries that light had to be a wave—until Einstein came along and once again kissed to life Newton's conviction that light also had definite particle properties. Twenty years later, people discovered that all particles—electrons, protons, neutrons, and so on—also exhibit interference effects. So if there is something about a light wave that acts like a particle, there is also something about particles that acts very much like waves.

As long as two such patterns repeat themselves with the same frequency, they will always interfere either constructively or destructively. Like two people marching at the rate of precisely ten steps per minute: if they start out in step, they will stay in step, and if they start *out* of step, they will stay out of step.

This is not, in general, the way things behave, however. Normally, two closely related patterns tend to get in and out of step, so that in some places they will interfere constructively, and in other places destructively. Periods of cooperation (in the case of

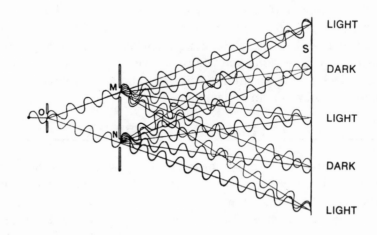

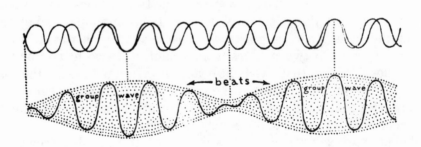

Many things can put two sets of waves into a pattern where they alternately get in and out of step—reflecting from thin films (like soap bubbles), bending around edges, or just having slightly different wavelengths. Periods of constructive interference alternate with periods of destructive interference.

people) alternate with periods of conflict; periods of loudness (in the case of sound) alternate with periods of silence; areas of brightness (in the case of light) alternate with areas of darkness. The result is the ups and downs of most relationships; the audible beats that tell when two instruments are slightly out of tune; the colorful bands in soap bubbles and oil slicks, opals and butterfly wings. The colors appear in places where light waves reflecting from two surfaces add up to one color but cancel another. Someone once described these colors as throbs of light.

Interference is a larger pattern that emerges from the superimposition of two other patterns. It therefore makes a very useful magnifier. The interference patterns that you see when two fences or fireplace grates overlap are much larger than the slats or wires themselves—just as the moiré patterns that adorn moiré silk and double layers of curtain are more readily visible than the silken threads that make them up. Interfering laser light is routinely used for surveying landscapes (even for measuring the distance to the moon), and the interference of X-ray light is used to study the nature of crystals. Recently, the interference patterns of radio signals from quasars billions of light-years away have been employed to measure the minute motions of continental drift.

Some people even describe the quantum states of the atom as a kind of interference pattern of particle waves. That is, where the particles (wavicles?) interfere constructively, it amounts to a stable state. Since all particles in nature are associated with wave patterns, all the atoms and molecules composed of them are patterns of such patterns. Perhaps physicist J. Robert Oppenheimer was thinking something along these lines when he said that the interference of electron waves gives rise to many "novel effects. . . . It is responsible for the permanent magnetism of magnets. It is responsible for the bonding of organic chemistry and for the very existence in any form that we can readily imagine of living matter and of life itself."

The fact that an electron is a wave, however, does not preclude its also being a pointlike particle—which all experimental evidence shows it to be. Matter waves are not *distribution* waves, that is, waves that are distributed over space. Rather, they are *probability* waves that chart the probability of a particle being in a certain place at a certain time. Like a wave moving through a row of dominoes or even a field of wheat, what is distributed in such a wave is not matter, but a state, or condition; a wave of information. This fits in with the forty-year-old observation of astronomer Sir James Jeans that electron waves are essentially "waves of

Interference patterns on a fireplace screen and supermarket cart. The interference patterns are magnified compared to the finer patterns of lines that make them up.

knowledge." The wave charts the probable place where a particle would appear *should we choose to measure it.*

When they're not interfering in each other's affairs (and even after they do), most waves move right through each other like so many ghosts. Sound waves, light waves, bow waves from boats, are continually crossing and interfering, yet surprisingly enough they arrive at their destinations essentially unchanged and intact; if they didn't, both visual and auditory images would be tangled in an impenetrable web of white noise. A strong wave in favor of a nuclear freeze and an even stronger buildup of new weapons pass through each other at the same time; so do right-to-abortion and right-to-life, the feminist mystique and the total woman, punk and prep. Art, science, politics, and economics are fairly rippling with contradictory trends that pass eerily through each other without influencing one another's far-reaching effects.

Yet this only seems strange until you consider that waves are not the series of short-term splashes they so often seem to be, but rather long-term carriers of *information.* When people view any kind of human concern—from civil rights to educational reform— as a momentary splash instead of an information-carrying wave, they ignore the lessons of history. They throw the baggage of the previous generations out—forgetting that it was built on pressures that had probably been accumulating for some time. Most important, they don't allow the wave to last long enough to gather the new inputs of energy it needs to keep going.

Unfortunately, watching for splashes at the expense of waves is all too prevalent. Our present fascination with the "new woman," for example, ignores the generations of mostly poor (and mostly black) women who worked outside the home while raising their children, just as our cyclically recurring worries about "the younger generation" ignore the fact that Socrates was saying in the fifth century B.C that "children nowadays are tyrants. They not only talk back to their parents, teachers and elders, but expect every luxury, gobble their food, chatter incessantly and sneer at any attempt to control them." On the other hand, I recently heard the editor of a major newspaper announce that he thought the subject of nuclear war was "old hat," and many science magazines dismiss quantum mechanics and relativity the same way. One hardly needs even mention politicians; every four (or two or six)

years, they present us with a splash of new policies—often before
the waves from the old policies have even had a chance to reach
their destined shores. (Even light waves take *time* to travel.) Isolationism or containment, confrontation or détente, economic stim-
ulation or deliberate stagnation—the various tunes tend to blend
discordantly into so much noise. No wonder people soon stop pay-
ing attention.

What makes waves different from splashes in the first place is
that all vibrating systems have a minimum of friction, a focused
and constant supply of energy, and—perhaps most important—a
built-in restoring force. That is, when they spring too far in one
direction (like a pendulum) they automatically are pulled back in
the other. That's what gives vibrations their springiness. But to
see a wave ebb and flow enough times to learn something about its
undercurrents of information, you have to be prepared to wait long
enough. Unfortunately, patience does not seem to be particularly
prominent in the inborn array of human qualities. One can't help
worrying that the world will end in a splash rather than a wave.

THE SHAPE OF THINGS

Not all waves are started by splashes, of course. Sand dunes
and snowdrifts and even the waves that flow through flags and
wheat fields are shaped by the force of wind. Light waves are
shaped by the ebbing of an electric field that creates a flowing magnetic field that, when it ebbs, creates another electric field, and so
on. Many ocean waves are propelled by the same pull of gravity
from the moon that creates the tides. The shape of waves—like the
shape of planets and soap bubbles, the twirl of a tornado and the
six-fingered star of the snowflake—all are sculpted by relatively
few forces and motions. And if some patterns seem to repeat themselves, it is because the forces that form them are strong currents
that flow throughout nature.

What forces mold the shape of waves, for example? Several
years ago I was asked to write what eventually amounted to a
small book on the subject of sine waves. Sine waves are the sort of
things that are music to physicists' ears, but I must admit they
held all the attraction for me of a high school calculus class on the
first warm afternoon of spring. It was almost a year before I found

a single nice thing to say about them. The year, however, was worth it. I learned that sine waves keep appearing as wiggles on teachers' blackboards because they appear so often in nature—from light waves to water waves. And the reason they appear so often is that they are the visible manifestations of a very fundamental kind of motion. It happens to be the motion of a pendulum, but it is also, a physicist told me, the basis behind a rabbit-hole system of transportation devised by Lewis Carroll while he was dreaming up *Alice in Wonderland*. It works like this:

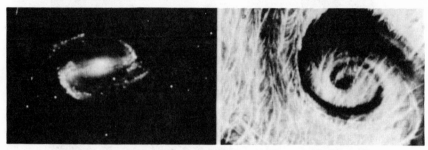

Which is a galaxy, and which is spilt milk? Patterns in nature tend to repeat themselves.

If you fell down a rabbit hole that went clear through the earth, you would be pulled toward the center by gravity until you reached a maximum speed of about five miles per second. Once you passed the midpoint, gravity would start pulling you backwards, but your own momentum would keep you going just about until you reached the opposite side—say Australia. If you forgot to climb out at Sydney, you would of course be pulled back again toward your starting point, and you could keep swinging back and forth like a human pendulum until friction slowed you down.

The beauty of the system would be that any rabbit-hole trip through the earth would take the same length of time—exactly forty-two minutes. Whether you jumped into an eight-thousand-mile rabbit hole to Sydney or a four-thousand-mile rabbit hole to Prague or a two-thousand-mile rabbit hole to Miami Beach, you would still arrive at your destination exactly forty-two minutes later. You would never get going quite as fast during the Miami trip because the acceleration due to gravity wouldn't be as great; but then again, you wouldn't have to go as far.

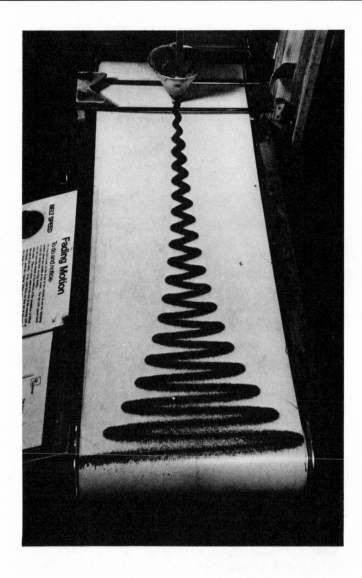

*As the sand-filled pendulum swings over the moving paper, it
draws a sine wave, which is the shape of Simple Harmonic
Motion. But there's another important pattern to see in the
sand waves: friction takes the same percentage of the remain-
ing energy away with each swing, just as inflation takes the
same percentage of your income with each passing year. The
result is an exponential curve—see "Small Differences."*

This explains why a pendulum (within limits) always takes the same period of time to swing back and forth even as the swings get smaller. The force that pulls the pendulum toward the center of the swing increases as the distance from the center increases; so as the distance gets larger, so does the force, and everything evens out in the end.

This principle is behind the precision of pendulum clocks, and also of quartz clocks, since quartz crystals vibrate with the same pattern of motion. So do flutes, violins, vocal chords, and atoms—which explains where the ubiquitous sine waves come from. A sine wave is a picture of this motion. That is, if you picture a pendulum with a paintbrush attached to its end, swinging so that the tip just brushes the top of a table, and you put a piece of paper under the swinging pendulum and pull it slowly toward you, the brush will paint a sine wave. And when a stone drops in water, it creates a wiggle with essentially the same properties—sending out water waves and sound waves (not to mention light waves), all in the shape of sine waves. Radio signals travel in sine waves because they are created by electric currents vibrating in much the same way.

Many things in the universe are round or roundish.

Waves, however, are just one of the common patterns that result from the relatively few forces and motions that mold nature.
Gravity is particularly powerful in this respect. Once Newton realized that it shaped the planetary orbits, "a lot of other things became clear," writes Feynman. "How the earth is round because
everything gets pulled in, and how it is not round because it is
spinning and the outside gets thrown out a little bit, and it bal-
ances; how the sun and the moon are round, and so on." Stars and
planets are round because gravity pulls matter toward other matter—a direction people on earth parochially call "down."

The parabolic path of falling water is shaped by the pull of
gravity.

The very shape of curved space is simply the pattern of the way
things "fall" under the influence of a gravitational field, in the same
way as iron filings fall into a certain shape when they come under
the influence of a magnet. The formulas that define these forces are
in one sense mathematical expressions of *behavior* patterns.

It is hard not to be impressed at how many things in the universe are round or roundish—just as it is hard not to be impressed at how many things travel in waves. Galaxies and weather patterns and DNA come in spirals—a kind of circle that grows or shrinks or moves along. The twirl of a tornado or the vortex formed by the water bubbling down the bathtub drain is molded by the interplay of inward pressure and outward momentum. Bubbles are built on surface tension. The parabolic paths of falling balls and water streaming from hoses and fountains are charted by the steady acceleration caused by gravity.

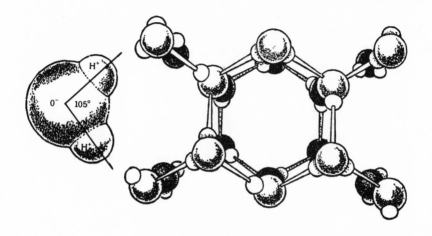

The airy symmetry of an ice crystal (right) reflects the arrangement of atoms in a water molecule (left).

Other forms are shaped by other forces—and often they have nothing to do with curves or waves. The airy symmetry of a snowflake is testimony to the strength and arrangement of water's hydrogen bonds. Trees, river deltas, blood vessels, and electric discharges branch in surprisingly similar ways, seeking sunlight or sea level or opposite charges. The size and shape of atoms is determined by the range of forces that hold their particles in place. Even the size of stars is controlled by a balance of forces. The inward pull of gravity is countered by the outward expansion

sparked by the heat of nuclear fires. Stars that are too small fail to
ignite, and remain "planets." Stars that are too big burn out
quickly (in star-time) in giant explosions. Thus the size of stable
stars remains within well-established limits. Ants, elephants, and
algae; honeycombs and trees; human bones and spiral horns; all
take shapes that fit the pull of nature's forces. In his famous book
On Growth and Form, D'Arcy Thompson said that any object was
essentially a "diagram of forces."

   Scientists look at patterns for clues to these underlying forces
—and so do parents, psychologists, economists, and an occasional
farsighted politician. The key is not to dismiss the pattern as something intangible and abstract, while seizing on the seemingly more
"concrete" individual cases. Patterns may seem ephemeral, but in
the end they are the enduring essence of things. They are the
waves of substance that linger long after the momentary splashes
of fate and fashion have gone silent.

14. CAUSE
AND EFFECT

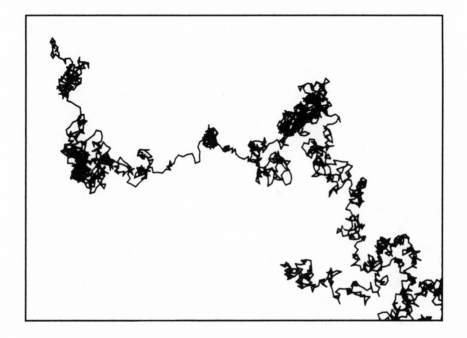

The Greek philosopher Democritus said he would rather understand one cause than be king of Persia. And no wonder: it is one thing to know *how* the universe works, and quite another to know *why* it works that way. People understand how gravity works, but not why, in the same way as they can explain to you how countries get into wars that wind up killing millions of their own people, but are at a loss to tell you why. One of the great unanswered whys in history concerns the unexplained astronomical amnesia that many historians say descended on the Western world between the time of the ancient Greeks and the Renais-

sance. Aristarchus was writing in the third century B.C. that the
spinning earth and other planets revolved around a central sun,
yet it was seventeen centuries later before this "discovery" was
resurrected by Copernicus and others. "We know how this happened," writes Arthur Koestler. "If we knew exactly why it happened, we would probably have the remedy to the ills of our
own time."

The search for underlying causes is innately and powerfully appealing—not the least because it implies the ability to control: if
you know what makes things happen (or not happen) you might be
able to make them happen (or not happen) again. Above all, people
like to think that there *are* causes. It is disconcerting to think that
events are random, that the things we see around us do not con-
form to an understandable (if still not understood) body of natural
law. Newton's ideas were readily accepted in part because they
contained a clearly applied formula for cause and effect: planets
orbited and falling objects accelerated in a certain way because
gravity pulled on them in a certain way. Darwin's ideas are much
harder for some to swallow—and at least some of this resistance is
rooted in the role his theory gives to random mutations. (Physicist
Sir John Herschel complained that Darwin's ideas about evolution
constituted little more than "the law of higgledy piggledy.") Even
Einstein refused to accept quantum mechanics because he thought
it did away with causality altogether. He wrote to his friend and
fellow physicist Max Born in 1944:

> You believe in the dice-playing god, and I in the perfect rule
> of law in a world of something objectively existing which I
> try to catch in a wildly speculative way. I hope that some-
> body will find a more realistic way, or a more tangible foun-
> dation for such a conception than that which is given to me.
> The great initial success of quantum theory cannot convert
> me to believe in that fundamental game of dice.

A strict belief in causality, on the other hand, completely contradicts another notion cherished by most people: the conviction
that they have free will. If every effect is produced by a cause that
itself is an effect produced by yet another cause, and so on in a
straight line that takes us back to the beginnings of the universe,

then everything we do must be predetermined—and must have been since the beginning of time. If the cause of your missing the train is a snowstorm that was caused by a warm front over the Atlantic Ocean two weeks ago, that itself was caused by a particular combination of winds and sunspots and so on, you could easily trace the "cause" back to any arbitrary event you wished— an event that would still have a cause, of course. Believing in this "endless chain of natural causes," argued Born, "leads necessarily to the idea that the world is an automaton of which we ourselves are only little cogwheels. . . . I cannot enlarge on the difficulties to which this idea leads if considered from the standpoint of ethical responsibility."

Newton's universe was such an endless chain of natural causes: the behavior of planets and presumably people could be calculated if only one had enough information and enough time. Everything boiled down to pieces of matter propelled and controlled by precise and predictable motions. Emotions and thoughts were merely manifestations of so much preset electronic circuitry. The illusion of "free will" was nothing more than the arrangement of atoms and molecules that made up the human body. "Exhorting a man to be moral or useful," said Sir James Jeans, "was like exhorting a clock to keep good time; even if it had a mind, its hands would not move as its mind wished, but as the already fixed arrangement of its weight and pendulum directed."

Still, quantum mechanics seemed to introduce an unacceptable degree of uncertainty and unpredictability into the universe. Perhaps not since Copernicus finally put the sun at the center of the solar system has science so profoundly ruffled people's philosophical feathers.

People talk about cause and effect, but rarely do they know if the effects are really caused by the causes. I read in the paper one day recently that the people of Nicaragua were starving, and that the cause was their socialist government. I read in the same paper on the same day that millions of people in the United States were going hungry, but nobody said anything about a socialist government. It reminded me of an article I read in *Discover* magazine about a solar eclipse: "According to Javanese legend," wrote author Dennis Overbye,

eclipses occur when the sun is swallowd by the severed head
of the evil giant Kala Rau, who was beheaded trying to
drink a magic potion of the gods. His body fell to the earth
and became the *lesung*, a concave vessel in which rice is
pounded to be hulled. During eclipses the Javanese beat on
the giant's body—their *lesungs*—to make the head let the
sun go. It always works.

Cause and effect.
It's all too easy to link two events in time and then say that one
was caused by the other according to one's own convenience.
Johannes Kepler's mother was arrested as a witch in part because
her visit to a neighbor's house unhappily coincided with the onset
of a major illness. No educated person today would say that beating on a *lesung* could cause an end to an eclipse. But you do hear
educated people say that welfare causes poverty, or that family
planning causes teenage pregnancies, or that the paucity of decent
television programming is caused by people's poor taste. Sometimes, seeking the cause of things is like asking which came first,
the chicken or the egg. Which is another way of asking: Did the
egg cause the chicken? Or the chicken the egg?
Even when one thing clearly and consistently follows another,
it is not always possible to make causal connections. Swallows flying low do not cause rain, any more than standing under certain
trees "causes" babies, or a clear night causes the moon to be full—
although there are superstitions to that effect. It is only the fact
that one thing almost always precedes another that gives it the
appearance of causality. As Born points out: "You can predict
(with the help of a railway timetable) the arrival at King's Cross of
the ten o'clock from Waverly; but you can hardly say that the timetable reveals a cause for this event."
Confusion about the relationship between cause and effect frequently leads to amusing (in hindsight) conclusions. For example,
Guy Murchie tells of a committee of scientists in the seventeenth
century who decided not to go ahead with a large investment in
Gutenberg's printing press because, they said, there could never
be a big enough demand for books. The reason? Only one percent
of the population could read. It probably never occurred to the
scientists that the availability of books could be a *cause* of people

learning to read, as well as vice versa. Today, this line of reasoning seems as silly as the argument that computers will never catch on because only a small percentage of people know how to use them.

Still, this very elusiveness of causes is what makes them such fertile territory for misinterpretation and superstition. The truth is that we rarely understand what forces are at work. What causes a family quarrel? Or a full-scale war? What causes a crime? Or a work of art? What causes two people to fall in love? Or to seek a divorce? Most of the time it's impossible to untangle the cause from the complex of circumstances that surrounds it. In many ways, the cause *is* that complex of circumstances—and it may include everything from the day's weather to the full weight of history. You might as well ask, What causes matter? Or what causes life? Or as Victor Weisskopf once said to me: "What is a cause? All you can say is, there are connections."

Connections between seemingly related events have been interpreted in many different ways throughout history. Ancient peoples didn't understand the connections behind spring and fall, or day and night, so they attributed these things (along with almost everything else) to the will of various gods. The gods were kept very busy, for there was a lot to do: not only floods and famines, but also the everyday events of everyone's life. Nothing happened as a result of consistent causes or understandable reasons. Even the planets would have stopped moving if the angels pushing them had stopped fluttering their wings. In Aristotle's day, it took no less than fifty-five separate spirits just to keep the seven known planets in motion.

Kepler, according to most sources, was the first person to seriously speculate in the sixteenth century that some "force" must be involved in the motions of the planets; he was the first to propose that things happen for rational, nonmystical reasons. Newton crystallized this kind of thinking in his vision of a great clockwork universe, where everything was controlled by forces. "For the eighteenth century, the world was a giant mechanism," wrote Robert Oppenheimer. All motions could be analyzed in terms of the forces producing them. In a sense, Newton substituted one kind of cause and effect for another. Forces took the place of spirits and gods. The world view of causes went from chaos to order, from total randomness to complete predictability.

Now it seems as though we have come full cycle. Quantum mechanics with its innate uncertainties has been accused of reopening the door to randomness, dispensing with order and causality, distilling the laws of nature into a kind of subjective mysticism. Or at least, so say some popular interpretations. In truth, all quantum mechanics has done is to bring to the fore a new *kind* of cause. And surprising though it may seem, there are many *kinds* of causes.

In the first place, there is the kind of cause that says things happen because it is the natural order of things. Copernicus, for one, thought that gravity was just a natural inclination of matter, bestowed by the Creator. Rocks "belonged" down, just as clouds "belonged" up. "The desire of every heavy body is that its center may be the center of the earth," wrote Leonardo da Vinci. Yet this kind of thinking goes clear back to Plato and Aristotle, who thought that slavery was the natural order of things just as they thought that circular motion was the only "natural" orbit for a planet. Aristotle's was a multilayered universe where everything had its proper place; a "cosmos graded like the Civil Service," Koestler called it. Aristotle's influence lasted a millennium and a half, and still persists with those who argue that some kinds of people are "naturally" poor or stupid, or that women "belong" at home. It is based on the same kind of thinking that says smoke has a "natural" tendency to rise or that the sun "belongs" in the center of the solar system. It is largely an admission that we do not know the forces or reasons involved. Like a child frustrated in her inability to make a reasoned case who finally argues, "Well, that's the way it *is!*"

Of course, our notions of what's natural do change—even in physics. Aristotle thought that the natural state of things was at rest and that you needed a force to keep them going; Newton said that the natural state of things could be in motion and that the "force" that kept them going was a "natural" property of all matter known as inertia.* Newton thought it "natural" that there was such a thing as absolute motion and absolute rest, but Einstein

*Newton's view was "righter" than Aristotle's mainly because it resulted in progress. See "Right and Wrong."

proved him wrong. Few people still view slavery as a natural part of the social order, but a surprising number of people still view Orientals as naturally cold, Latins as naturally slothful, and so on.

Natural causes of the sort that Aristotle had in mind are very different from the causes we associate with forces. A force requires an exchange of energy. You punch someone, and he falls down. You blow on a birthday candle, and it goes out. You wave a greeting and send out a subtle influx of air molecules into the surrounding space. Gravity pulls and an apple falls to the ground. And yet, even gravity remains essentially a name for a pattern that is associated with a still-not-very-well-understood behavior of objects. Newton never claimed to understand why gravity worked, or how it spread through space. "I do not deal in conjectures," he said. Or as Richard Feynman put it: "At the time of Kepler, some people answered this problem [of what makes the planets go around the sun] by saying that there were angels behind them beating their wings and pushing the planets around an orbit. As you will see, the answer is not very far from the truth. The only difference is that the angels sit in a different direction and their wings push inwards." (The inward push of the angels, of course, is gravity.)

The notion that forces are carried by fields, or exert their influence on bodies by means of fields, seems uncannily connected to Aristotle's notion that everything has its proper place in the cosmos. That is, the apple falls not because gravity pulls on it but because the apple naturally falls to its proper place in the gravitational field, just as iron filings naturally fall into place around a magnet. In fact, the tendency to fall into a natural state is common to all things in nature. Not only do apples fall to their ground states, but so do atoms—in the process, giving off energy and emitting light. All things seek their lowest energy state just as water seeks its lowest level. And this tendency to seek the lowest or most stable level is considered a perfectly legitimate *cause* of things.

Yet there are two important differences between this kind of cause and the causes that operated in Aristotle's universe. In the first place, when an apple or an atom changes its state, an exchange of energy (or force) is involved. It takes energy for the tree to raise the apple above the ground, and the apple returns some of that energy to the ground (in the form of heat) when it finally falls. In the same way, it takes energy to excite an atom into an "unnaturally" high state, and the same atom releases energy when it "falls" back to the ground state.

Even more important, the natural states of apples and atoms are highly regular and grounded—so to speak—in a vast network of empirical evidence and theoretical understanding. They are connected to many other well-understood and well-documented phenomena. They are consistent, not capricious. Gravitational fields behave the same way toward men as women, apples as oranges. Atoms change states whether or not they have made penance to a particular god. Forces and fields do not discriminate. They treat the atoms in earth, fire, air, and water equally.

This relates, in a way, to a particularly peculiar kind of "cause" —symmetry. All the strange effects of relativity—from time dilation to curved space—result from the idea that the laws of nature are symmetrical; that it does not matter whether you are moving or at rest; that the speed of light is always the same; that there is no absolute rest frame in the universe. What is the "cause" that makes time slow down as you travel at greater speeds? Why do electrons traveling at 99.9999 percent the speed of light in giant accelerators gain 40,000 times their normal weight, or mass? The "cause" is symmetry: the fact that the speed of light is measured the same no matter how fast you are moving; that some differences do not make a difference. The reason that rulers can change their dimensions and clocks can tick off different times is that you always see the *same* laws at work in the universe—no matter how you are moving about in it.*

Like other causes, the idea that symmetry can be a shaper of things goes back to ancient Greece. Plato said that the shape of the world must be a perfect sphere, and that all the planets must travel in perfect circles, because only circles are perfectly symmetrical. Circular motion has no beginning and no end. It doesn't matter how you look at a circle, it always looks the same. Some people even argued that objects "gravitated" toward the center of the earth because that made things nice and symmetrical. The notion of symmetry was such an appealing "cause" that it wasn't until

*See "Relatively Speaking."

Kepler that people officially recognized that the planets orbited not in circles at all, but rather in ellipses. My friend the physicist likes to point out that symmetry is also the basic argument behind social ideas such as the need for civil rights legislation: people are innately symmetrical before the law. You can't treat blacks and whites (or men and women) differently because their fundamental needs and abilities are the same.* In fact, while it may be easier to think of causes in terms of physical forces ("after all," says Jeans, "our hairy ancestors had to think more about muscular force than about perfect circles or geodesics"), no concept of cause that relies on force can be entirely subjective. According to relativity theory, force itself depends on the observer's motion and point of view. So causes interpreted as forces must always be necessarily somewhat subjective, too. Looking for causes in symmetries leads to much clearer and more uniform conclusions.†

Some people even think that unity, or harmony, or beauty, can be a cause. According to Pythagoras, the configuration of the heavens was "caused" by the requirements of musical harmony: the spacing between the planets corresponded to harmonic intervals, planned so they would continually play the "music of the spheres." Present-day scientists are often guided in their work by the search for a similar underlying harmony, or at least a unity among the various disconnected parts of nature.‡

Only in quantum mechanics—in the physics of subatomic things—do causes seem to spring out of nowhere, or at least out of chaos, which is much the same thing. Even the whims of the ancient gods seemed more understandable than the inner workings of the atom. Yet there is a magical order to the seeming randomness of atomic events. And the very reality of the order in random events has totally altered the meaning of our notion of *causes*.

What is the "cause," after all, that makes a tossed coin turn up heads fifty percent of the time and tails fifty percent of the time? What is the cause that makes a certain number of radioactive atoms decay now instead of later? What is the cause that determines how many times the roulette wheel will come up red or

*See "Shadows and Symmetries."
†See "Relatively Speaking."
‡See "The Scientific Aesthetic."

black? It is the same kind of cause that keeps an egg from unscrambling itself, yet makes it almost certain that a room will unclean itself; the same cause that makes the heat flow from the warm drink into the melting ice cube and not vice versa. The cause that makes these things happen is simply that they are more likely to happen than the alternative. A cause, my friend the physicist says, is anything that makes something happen at a slightly higher rate. Those things that happen more often are those things that have

The Probability Machine at the Oregon Museum of Science and Industry (OMSI). A lot of random events can add up to a predictable pattern.

the most ways of happening, which is another way of saying that most probable things happen more often. So probabilities can also be causes. This seems completely nonsensical only until you stop to consider the evidence.

Take a bunch of billiard balls, for example. (Physicists are always taking a bunch of billiard balls.) Many science museums have an exhibit where a bunch of billiard balls are allowed to fall at random through a forest of pegs sticking out from a wall. The balls hit the pegs and bounce this way and that—all at random. And yet when they all collect at the bottom in a bin, they take the shape of a surprisingly predictable curve. People like to try this experiment again and again, precisely because the result seems so unlikely. How do you get that nice pattern out of all that random motion? What causes it?

Or take a drunk leaning on a lamppost. Say he decides to take a walk. First he steps forward, then teeters sideways, then stumbles backwards, then goes off in another direction—all at random. Can you predict how far the drunk will travel after taking a given number of steps? Incredibly, it turns out that you can. You can even describe it in an equation. The total distance traveled equals

There is actually an equation that predicts how far someone (or something) will travel during a walk composed of completely random steps. This random walk was generated by Ron Hipschman's computer.

the average length of each straight step, times the square root of the number of steps he takes. So if his average step is one yard

long, he'll travel ten yards in one hundred steps. (You only don't
know which *direction* the steps will take him.)

Using this same kind of analysis, Einstein calculated the size of
molecules from looking at the chaos of microscopic collisions known
as Brownian motion. Small particles suspended in a liquid—say,
plant spores or oil droplets—are buffeted about at random by the
unseen molecules of the liquid. The result is that they move about
somewhat like drunks—and their paths can be calculated in much
the same way. The same equation is used to predict how fast
smoke—or pollution—will spread in the sky.

Isaac Asimov even got the idea for his *Foundation* series of
science fiction stories from the statistical order that emerges from
the random motions of gas molecules. Given quintillions of quintillions of molecules, he says, you can predict exactly how the sample will act.

> The motion of any one atom or molecule is completely un-
> predictable—you can't tell where, in which direction, or
> how fast it will move—but you can average all the motions
> and from this deduce the gas laws. It occurred to me that
> this might also be true for human beings. A human by him-
> self is quite unpredictable, but a mob is usually a little more
> predictable. What if we applied a kinetic theory of humans
> to some future time, when there were millions of planets full
> of people?

As Asimov points out, this is only science fiction. People are
much more complicated than gas molecules. Yet the idea of statistical probability takes on a strange concreteness when it comes to
subatomic particles: all particles can also be described as waves—
this is part of the dual nature of matter. But the particles can also
be pointlike, because the waves are not matter waves; they are
probability waves. The wave charts the probability that the particle will be in a particular place at a particular time if you should
choose to measure it. It may be a little misleading to call it a probability wave, because it's not the probability that's being propagated. It's the light or the electric field or whatever property makes
up the particle that's being propagated. A particle wave tells you
the same thing that the curve produced by the billiard balls falling

through pegs on a wall tells you. The billiard balls themselves do not curve; only their distribution does.

Are these particle waves real things? Nobel Prize winner Born,* among others, says yes—primarily because probability waves share the same properties of other things we consider "real." Their existence is invariant (it does not vary under different conditions) and highly predictable. They can be relied upon as much as you can rely upon the fact that your next step will fall on solid ground. "I personally like to regard a probability wave . . . as a real thing," he writes. "How could we rely on probability predictions if by this notion we do not refer to something real and objective?"

People often make the mistake of dismissing probabilities as mere abstractions, and therefore unreal. They do not take them seriously as causes. Yet the high probability of, say, a natural disaster or nuclear war can cause it to happen as much as a high probability causes a coin to land on its tail fifty percent of the time. Probable causes, like probability waves, are real because they *work*. The laws of statistics are laws of nature like any others. Probability is as much a shaper of things as gravity.

In exploring the roots of causes, however, you inevitably find yourself stumbling over some serious paradoxes. In the first place, it may look as if different laws of nature apply to small numbers of things than apply to large numbers of things. The behavior of one coin or atom is completely unpredictable, while the behavior of hundreds of coins or trillions of atoms is quite precisely predictable. Unpredictability implies randomness, which is equated with lack of cause. Either something happens on purpose (with a cause) or by accident (at random); you can't have it both ways. The behavior of atoms, according to this interpretation, is random and therefore acausal.

What causes a radioactive atom to decay, for example? Say you

---

*Theoretical physicist Max Born won the Nobel Prize in 1954 for his work on statistical quantum mechanics; he was a pioneer in trying to interpret the meaning of quantum theory into everyday language, and so—according to people like Victor Weisskopf—may have overstated the case. Weisskopf would say that a particle wave isn't "real" so much as it is a clumsy approximation of reality—the clumsiness being rooted in the inappropriateness of our language for dealing with quantum mechanical things.

take a milligram of radium. You can predict fairly precisely how many atoms will disintegrate with every passing second. Yet there is absolutely nothing you can do to change this situation. The rate of decay is not affected by anything in the environment. You can make it hotter or colder. You can alter the motion, or crowd the atoms closer together, and the rate will stay the same. On the other hand, nothing in the past history of the atoms determines what they will do, either. Any milligram of radium from anywhere in the universe will behave exactly the same. There is no way internally or externally to change the situation. There is no determining factor in the past or in the present. Therefore radioactive decay seems truly to be an event without a cause. And, in fact, almost anything that has to do with a single atom exhibits this same "acausal" property.

Yet "how could this be," asks physicist John Wheeler, and "leave the largely familiar world intact as we know it? Large bodies are, of course, made up of atoms. How could causality for bullets and machines and planets come out of acausal atomic behavior? How could trajectories, orbits, velocities, accelerations, and positions re-emerge from this strange talk of states, transitions, and probabilities?"

If God plays dice with the universe, then he (or she) presumably plays a great deal of dice; otherwise, how to account for the familiar and predictable laws of nature?

In the second place, we are left with the further and perhaps more fundamental paradox that *chance* follows *laws*! And that events ruled by cause and effect, on the other hand, are seldom precisely predictable! It seems, as Born writes, "a hopeless tangle of ideas." Fortunately, Born* not only spells out the nature of the tangle; he also proceeds to *un*tangle it. And the route he takes was somewhat familiar to me because it echoed a favorite theme of my friend the physicist: the often misinterpreted role of prediction in scientific understanding.

Scientists and science writers are forever saying that subatomic events are unpredictable, and therefore random, and therefore acausal. But Born points out that causality and determinism

---

* All this and much more is from Max Born's excellent book *The Natural Philosophy of Cause and Chance.*

(or the ability to predict) are two different things. Causality means that one thing depends on another thing. (As Weisskopf would say, "there are connections.") Determinism means that you can predict the future based on this causal relationship. But Born, among others, says no. Just because you cannot predict whether a tossed coin will turn up heads or tails does not mean the event is acausal. There are forces at work that determine whether the coin will turn up heads or tails. But you cannot measure them well enough to predict the outcome of the toss—or at least not without interfering with the results of the measurement itself.

A snippet from *Scientific American*'s "Fifty Years Ago" column for October 1933 gave much the same interpretation: "It was discovered a few years ago that things as small as electrons do not individually obey the law of cause and effect," the writer began. ". . . Right here a number of thinkers made false deductions and, as A. S. Eddington put it, 'science went off the gold standard.' What these thinkers failed to grasp was that mere indeterminability does not in itself establish indeterminacy. A thing may be indeterminable but not indeterminate. Nature knows what she is doing, and does it, even when we cannot find out."

Our ability (or inability) to predict something does not necessarily depend on our understanding of causes—something that becomes self-evident when you consider how many things we can predict even though we do not understand them (swallows flying low predicts rain), and the many other things we cannot predict even though we *do* understand them. Take weather, for example. The forces behind weather are well understood. But weather itself is highly unpredictable—primarily because it is so complicated. You probably can understand and predict the course of a single air molecule as it responds to changes in humidity and atmospheric pressure. But take a whole bunch of air molecules, and you are lost.

As in foreign affairs and personal affairs, one small unseen effect can be enough to change the entire configuration of events. Nature (including human nature) is often too complicated and interconnected for neat categorizations into "cause" and "effect."

So did the innate uncertainty in quantum mechanics do away with causality or didn't it? Is the universe at its core a precisely tuned clockwork mechanism? Or is it random—a room full of tossed coins? Summing up the fruits of quantum physics, Robert Oppenheimer wrote: "We saw in the very heart of the physical world an end of that complete causality which had seemed so inherent a feature of Newtonian physics." But Max Born concludes: "The statement, frequently made, that modern physics has given up causality is entirely unfounded. . . . Scientific work will always be the search for causal interdependence of phenomena."

It may be that the only thing that has been lost by opening the paradoxical Pandora's box of quantum physics is the assumption that understanding causes also means the ability to predict and control things. As Weisskopf points out, you still know that a radioactive atom will decay, and you still know how it will decay— "you only don't know when." In fact, the crux of the uncertainty principle boils down to a matter of timing. Because the more accurately you try to determine exactly *when* something will happen, the more you make other factors obscure. On the other hand, there is a sense in which time itself becomes a cause of things. "Give me a million years" writes Stephen Jay Gould, "and I'll flip a hundred heads in a row more than once." When it comes to evolution, "time is in fact the hero of the plot. Given two billion years or so, the impossible becomes possible, the possible probable, and the probable virtually certain."

You don't need to know every factor in the causal equation, that is, to determine the probable outcome. You know that given a certain number of handguns in general circulation, a given number of people will be killed. You only don't know who. You know that when prepared infant formula is widely marketed in underdeveloped countries, babies die—just as you know that reducing the highway speed limit to fifty-five miles per hour saves tens of thousands of lives. Clearly you can't argue that infant formula kills babies or that high speed kills people by traditional notions of cause and effect. But you can certainly say that "there are connections."

In the end, Born harks back to the idea of complementarity. Strict causality and absolute randomness both have their place in the scheme of things, but both together are as inconsistent as images of waves and particles. In fact, if you take both arguments to their logical (linear) conclusions, they make no sense at all. For if you say that any kind of cause at all determines the way things (or people) behave, then you have to come to the conclusion that everything is predetermined. On the other hand, if you say that nothing determines the way things behave, then you must conclude that everything is random. If there are causes for everything, then we are cogs in the clockwork. If there are not causes, then we are so many dice.

On the surface, causality and randomness may seem to be mutually exclusive; but on closer inspection, they must be seen as complementary facets of a larger reality.

15. ORDER AND DISORDER

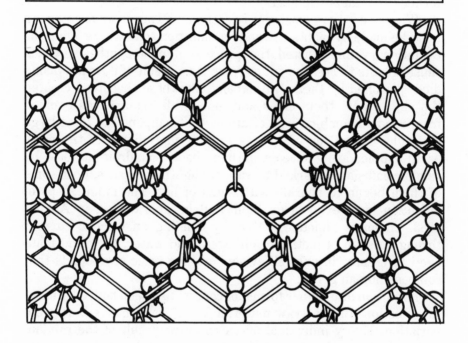

For all the seeming order in the natural world—the crystals and snowflakes, light waves and spiral galaxies, tightly organized ant colonies and neat, elliptical planetary orbits—much of what we see around us is ruled by randomness. Chance is not only a source of statistical causes,* it also accounts for most of the familiar forms that inhabit the universe, including ourselves. The shapes and sizes of stars are determined by the interplay between electrical pressure and gravity, but just why a certain star should

---

*See "Cause and Effect."

happen to be born in just such a place at just such a time is all a matter of chance: a small random fluctuation in the configuration of molecules floating around in space can cause a few of them to clump together, making gravity a little stronger in that corner of the vacuum; more molecules and particles are attracted to this place, making the pull of gravity even stronger; eventually, even such a tiny chance trigger can turn out to be the underlying force behind the formation of an entire galaxy.

We owe our very existence to accidents. The air we breathe is a mistake made by plants. A random mutation somewhere along the evolutionary line started the process of photosynthesis, which causes plants to breathe out oxygen. As a result of it, many plants died, poisoned by their own exhalations; those that lived created the sky. Plants themselves and indeed all living forms evolved from random combinations of atoms that happened to form large molecules suited for life; these organic ingredients of the early earth bubbled away for eons in their chaotic cauldron, buffeted by random collisions, sparked by random lightning flashes, until some of them began to sparkle with signs of life. The infant steps of evolution have even been duplicated in laboratories: in August 1983, chemists announced that by sending an electric current through a soup of methane, nitrogen, and water, they had managed to create "in one fell swoop" all five basic compounds that compose human genes.

Evolution into higher forms, of course, has been honed by natural selection. But the raw material for all such change is random variation. Every individual on earth is the result of the random mating of his or her parents, and countless other unaccountable events. "Whenever an infant is born," writes naturalist Loren Eiseley, "the dice, in the shape of genes and enzymes and the intangibles of chance environment, are being rolled again. . . . Each one of us is a statistical impossibility around which hover a million other lives that were never destined to be born—but who, nevertheless, are being unmanifest, a lurking potential in the dark storehouse of the void."

But what do we mean when we say that events are random? That things are ordered or disordered? That our lives are ruled by chance? As Guy Murchie and others point out, we use the term "chance" in several different, and quite contradictory, ways. We

say that something happens by chance in the sense of luck, or accident. This kind of chance is entirely unpredictable. On the other
hand, we also use chance to mean probability—a way of predicting
something. We predict a forty percent "chance" of rain, or the odds
of drawing a straight flush.

*Crystals are orderly because they look different depending on
your point of view. If they melted into a liquid, the way you
viewed them would not make much of a difference.*

The meaning of order and disorder can be similarly confusing.
What would you call a situation, for example, in which everything
looked the same no matter which angle you viewed it from? In
which everything was distributed evenly? Which was entirely undifferentiated and homogeneous? A room full of coins that had exactly as many heads on the left half of the floor as on the right half
of the floor? Or a universe where all the fundamental forces were
the same and operated at the same strengths and distances? Or a

classroom in which no distinction was made between the boys and the girls? Such a situation, it turns out, is both highly symmetrical and highly *disordered*. It is as if I took the entire contents of my file cabinet and flung them on the floor, so there was no longer any difference between "Forces" and "Receipts," or "Gravity" and "Weisskopf." It is completely democratic, but unspecialized.

Order, on the other hand, is much more authoritarian. A closet is well ordered if the shoes are separated from the shirts, and the skirts from the trousers. An army is well ordered if the privates are distinct from the colonels, just as an ant colony's order is based on the specialization of its member parts. To bring order to a classroom, you could separate the boys' activities from the girls'—or you could make sure that every student did the same work at the same time. This might seem at first like bringing more homogeneity, or sameness, into the situation. But in fact, by separating the math from the spelling and the science from the reading, you would be increasing the amount of heterogeneity, or differentness. Orderly situations are well differentiated. Today's universe with its four fundamental forces—with its atoms and organisms and galaxies—is far more ordered than the early universe, which was little more than a hot mass of homogeneous soup. Aristotle's universe was also highly ordered: everyone from slaves to shoemakers, and everything from rocks to planets, had a proper and permanent place.

Order is obviously a lot more complex than disorder—something which makes chaos by comparison seem simple! A species is well ordered if its eating functions are well separated from its eliminating functions, for example, or if it has many other well-specialized parts. Curiously, we say we are a "higher order" of species, meaning that we are more complex and therefore clearly better than other species. But Darwin himself was careful not to attribute terms like "higher" and "lower" to the fruits of evolution. "For if an amoeba is as well adapted to its environment as we are to ours," writes Stephen Jay Gould, "who is to say that we are higher creatures? . . . Hair on a mammoth is not progressive in any cosmic sense." Only if it begins to get colder.

Curiously, too, a world in which all countries shared the same amount of resources, or in which all people shared the same amount of power, would be highly disordered, in the sense of lacking a strict hierarchical organization. But equality is not inevitably disorderly. Crystals manage to make order out of chaos from the ground up, without relying on "leaders." A homogeneous situation always implies a certain lack of distinction, of course, a loss of identity. Yet it is meaningless to apply value judgments to terms such as "order" and "disorder." People often automatically equate "disorder" with "bad," even though disorder is a measure of (among other things) warmth. Others are repelled by the idea that the raw material for life was created randomly, but as Gould asks: "Shall we appreciate any less the beauty of nature because its harmony is unplanned?" You cannot assume that an organization is well run just because it is highly specialized; dinosaurs were highly specialized, too. On the other hand, none of us would be here if there weren't an amazing amount of order in the universe. Even what we call "mess" or "dirt," for that matter, is open to interpretation: as Murchie points out, a bread crumb becomes "dirt" only when it falls to the floor. The composition of the bread crumb stays the same, just as atoms and molecules are the same whether they are ingredients in a "mess" or in a work of art. As the old Serbian proverb put it: "Be humble for you are made of dung. Be noble for you are made of stars."

Be that as it may, most people seem to be smitten with an innate love of orderly things—bureaucracies, lists, snowflakes, and well-ordered closets. Order, in nature, is beauty. The physicists' perennial search for a Grand Unified Theory that links the fundamental forces of nature is at least in part aesthetically motivated—a search for order. Einstein spent most of his life vainly looking for it. Recently, some physicists think they have gotten a clue to how the strong force might fit in with the now unified "electro-weak" force. But it drives them crazy that gravity eludes all efforts at unification. Or as Caltech president Marvin Goldberger once said to me: "It would be just obscene if everything else fit together and gravity was sitting out there like a big bird."

Physicists (as one might have suspected) have even made orderly studies of disorders. Not only can they predict at least some aspects of random events; they have also discovered basic laws of nature in disorder. In fact, there is something eerily orderly about disorder. It is measurable, and even largely predictable. It is the one quantity in the whole universe that always, inevitably, increases. Like death and taxes, one thing of which you can always be certain is an increasing amount of large-scale disorder.

You obviously don't need to be a physicist to understand this. It came home to me (so to speak) one day last year as I was sitting in my kitchen contemplating all the freshly frozen food that was going to spoil because my refrigerator had just broken. This came on top of the discovery that my right rear tooth needed root-canal work and that my son needed new sneakers. The garden was turning to weed and my hair was turning gray. The house needed paint and the typewriter needed repairing. My best sweater was developing holes, and I was developing a deep sense of futility. After all, what was the point of spending half of Saturday at the Laundromat if the clothes were dirty all over again the following Friday?

Disorder, alas, is the natural order of things in the universe. It is measured by a property physicists call "entropy," and the fact that it always increases emerges from the Second Law of Thermodynamics. One way of stating it is this: "Natural processes tend to proceed toward a state of greater disorder."* Most fundamental physical quantities—such as energy, matter, momentum, spin— are conserved. That is, you get out exactly what you put in, and the amount present in the universe always stays the same. You can't get rid of energy any more than you can create it out of whole cloth—you can only change it from one form to another. (M.I.T.'s Philip Morrison likes to chide people who talk about "conserving" energy, because it is always conserved anyway. What they mean is keeping it in a usable form.)

But entropy is another thing. You always get *more* than you started with. Once it has been created, it can never be destroyed. The process is not reversible. The road to disorder is a one-way street. (The good news is that you can borrow energy from one part of the universe to create order in another part of the universe, thereby creating "islands of order," like stars and people. But more on that later.)

Because of its unnerving irreversibility, entropy is often called the arrow of time. Everyone understands this instinctively. Chil-

*The Ideas of Physics, by Douglas C. Giancoli (Harcourt Brace Jovanovich, 1978).

dren's rooms, left on their own, tend to get messy, not neat. Wood rots, metal rusts, people wrinkle, and flowers wither. Even mountains wear down; and even atoms can decay. In the city, you can see entropy in the run-down subways and worn-out sidewalks and

Increasing disorder is the arrow of time. If this photograph were a film, you would know without question whether it was running "forward" or "backward."

torn-down buildings and fallen bridges. You know, without asking, what is "old." If you were suddenly to see the paint jump back on an old building, you would know that something was terribly wrong. An egg does not unscramble itself any more than Humpty Dumpty could put himself back together again.

But what is the cause of inevitably increasing entropy? What prevents Humpty Dumpty from spontaneously reassembling? Or for that matter, people from growing younger instead of older? The answer is essentially *probability:* the combined results of countless random events. "Irreversibility," Richard Feynman concludes, "is caused by the general accidents of life."

Take the air in my kitchen just before the refrigerator broke. It was a highly ordered situation—a low degree of entropy. All the cold air was kept inside the refrigerator, and the warmer air was isolated outside. The minute the machine stopped working, however, the cold- and warm-air molecules were free to exchange energy at random. As they were jostled about this way and that, it was possible, of course, that all the cold (meaning slow-moving) molecules would be bumped back in the direction of the refrigerator. But it would be highly unlikely. The likely result was that the cold and warm molecules would wind up randomly mixed, and that I would be left with a lukewarm mess.

Of course, there was nothing to prevent any one molecule from moving one way or another. No force pushed the cooler molecules away from the refrigerator. In fact, any one of the slow-moving molecules stood about as much chance of being bounced back toward the refrigerator as of being bounced away from it. But take trillions and trillions of warm and cold molecules mixed together, and the chances that all the cold ones will wander toward the refrigerator and all the warm ones will wander away from it are practically nil.

Entropy wins not because order is impossible, but because there are always so many more paths toward disorder than toward order. There are so many more different ways to do a sloppy job than a good one, so many more ways to make a mess than to clean it up. If I put a baby in front of a typewriter, the odds are roughly one in forty-three that she will type the letter *a*. There is less than a chance in a million, however, that she could consecutively strike the letters that would spell out Shakespeare. And the chance is so infinitesimal that she would type the complete works of Shakespeare that we call it impossible.

This is precisely the same reason that it is "impossible" for warm air to randomly excite the molecules in a melted stick of butter so that they would regroup spontaneously into a bar—or that ice cubes would spontaneously appear in a lukewarm drink.

Because there are far more opportunities for things to turn out otherwise. A baby learns to take the puzzle apart long before he learns to put it back together again because there are so many more ways for the puzzle to come apart than there are for it to go together. There is only one possible way for Humpty Dumpty to remain a recognizable whole, but there are an infinite number of ways for him to fall to pieces. In fact, this explains why most random mutations are harmful rather than helpful: there are simply many more ways that a random change in the nature of things would turn out for the worst.

Of course, the more pieces there are in the baby's puzzle, the harder it is to put it back together again. In a sense, entropy boils down to the number of possibilities. A coin can only come up heads or tails. But a dust particle in a room can take an almost infinite number of possible positions, and so add up to a far "messier" situation. If my kitchen contained only a dozen or so air molecules, then it would be likely—if I waited a year or so—that at some point the six coldest ones would congregate in the freezer. But the more molecules in the kitchen—the more factors in the equation—the less likely it is that their paths would coincide in an orderly way. Or as one physicist put it, "Irreversibility is the price we pay for complexity."

It is not too farfetched, I think, to make an analogy to the increasing disorder most people sense in modern society. Everything from human relations and social institutions to law and medicine are more complicated than ever. Everyone has more options than ever—which also means more ways to make mistakes, and more power to make big ones. It is easy to rebuild a house, but extremely difficult to rebuild a city or a society. As you walk through the streets of Manhattan, for example, it is impossible not to be impressed by the sheer amount of activity going on, the numbers of people, the variety of endeavors. Is it any real wonder that things seem to be falling apart? In the end, does it matter?

It turns out that it does matter, mainly for the reasons that got scientists so interested in entropy in the first place. And that had to do with the amount of useful energy available in machines. The Second Law of Thermodynamics also says that no matter how efficient you make a machine, you always get out less useful energy than you put in. The excess energy gets dissipated into heat, or entropy. And that energy can never be retrieved. For while energy itself is always conserved, *usable* energy is not. Once the water has flowed over the falls, it has lost its potential for useful work. Once the cold-air molecules had been let out of my freezer, they lost their potential to perform a purpose; they allowed my butter to melt, my milk to spoil, my frozen vegetables to decay. "What has been 'lost' in the irreversible process is not *energy*," says Feynman, "but *opportunity*."

Entropy implies a useless form of energy, a kind of random agitation that we associate with loss of purpose. The nice part about entropy is that everything evens out in the end; entropy keeps us warm. But the bad part of entropy is that it means opportunities to accomplish things are inevitably lost. So it turns out that it does matter very much just how "disorderly" things are.

There is an inherent contradiction in all this, however. On the one hand, physicists say that an increase in disorder is inevitable—and you can see the results of this around you wherever you look. On the other hand, the universe is clearly an increasingly structured place—and the results of this are obvious too. The primordial ball of fire has cooled and condensed from a hot, amorphous mass into elements, stars, planets, and people. What we see in the universe is increasing order, not disorder.

The paradox becomes compounded when you consider the very close relationship between disorder and temperature. Heat is a measure of random motion. Warmth (like warm, melting butter) means disorder. The early universe was probably a disorderly hot mess of particles and radiation. Today even the stars are made not of atoms, but of something more like an undifferentiated stew of atomic particles—pieces of atoms torn apart by the high energies inherent in extremely hot temperatures. This state of matter is called a "plasma," which means "mixture," and it accounts for all but a fraction of the matter in the universe.

Cold, on the other hand, is associated with order. Only when temperatures cool considerably can stable atoms form out of atomic constituents like protons, neutrons, and electrons. Only when they cool further still can more complex molecules form from the atoms. A molecule of water breaks down into hydrogen and oxygen atoms if things get too hot. Steam is less orderly than water, and water is less orderly than ice. All the familiar "orderly" states of water—ice cubes, snowflakes, hailstones, crystals—can form only at comparatively low temperatures. The increasing order in the universe is also associated with a cosmic cooling down. The trillion-plus-degree temperatures present at the creation—the so-called Big Bang—are now measured at a mere three degrees above absolute zero, about minus 450 degrees Fahrenheit. Physicists say that the four fundamental forces have somehow "frozen" into their now separate states.

But how can the universe be cooling down and becoming more ordered when entropy is universally increasing?

The answer is simply that you pay for order with energy. It takes energy to produce a neat set of files or a well-ordered closet, just as it takes energy to produce an atom or a star or to keep all the cold-air molecules inside the refrigerator. It *is* possible to create order in the universe. But the energy to do it must be borrowed from other parts of the universe. The islands of order that bask in the chaos—the crystals and snowflakes, the buildings and cities—all exist at the expense of something else. We get the energy to construct buildings, for example, primarily from the fossil fuels needed to fire the steel mills and drive the cranes and trucks; in doing so we increase the familiar form of entropy known as smog. The price of creating order in any one corner of the universe is increasing disorder somewhere else. (This is not so different, perhaps, from the way foreign policies designed to bring order to the political affairs of one country can so often wreak havoc in other parts of the world.)

When it comes to the whole universe, however, the increase in disorder simply dissipates into the vastness of infinite space. After all, if entropy can be measured as the number of possibilities, then the number of possibilities in infinite space is clearly limitless! The entropy goes into radiation that is simply lost in that big empty space.

The most obvious exception to entropy is life. A seed soaks up some soil and some carbon and some sunshine and arranges it into a rose. A seed in the womb takes some oxygen and pizza and milk and transforms it into a baby. Death is an extreme form of entropy. Life is the epitome of order, of purpose incarnate. To live is to be in constant battle with the Second Law of Thermodynamics, and yet it is a battle more often won, it seems, than lost. Despite all odds, flowers bloom in the desert and children bloom in slums. "If something or somebody has a will to live, you see," says Murchie, "it or he [or she] must resist diffusion and move from disorder to order, which means avoiding all those easy paths away from the previous position in favor of returning to it, or, better, staying with it from the beginning. Basically it connotes sticking around."

Of course, it takes a lot of energy to produce a baby, just as it takes a lot of energy to make a tree. And it is that expenditure of energy that makes life precious. The ingredients are only the same everyday atoms that make up rocks and water and air; the stuff of nonliving things is exactly the same as the stuff of living things. The difference is in the arrangement of atoms; in the kind and the degree of order. If the ingredients of life are cheap (ninety-eight cents, or ten dollars, or whatever), the final composition is not. Carl Sagan talks about a certain Harold Morowitz who reportedly calculated what it would cost to manufacture with current technology all the molecules that it takes to make a person. And it turns out that merely to make the molecules—never mind a cell or a liver—would cost more than ten million dollars.

Though combating entropy is possible, then, it also has its price. That's why it seems so hard to get things together, and so easy to let them fall apart. The inevitable accidents and obstacles of life almost guarantee that things will get off track, bounced onto random paths. Disorder is the path of least resistance—the easy, but not the inevitable road. Social institutions—like atoms and stars—decay if energy is not expended to keep them ordered. Friendships and families and economies and efforts at international peacemaking all fall apart unless a constant effort is made to keep them working and well oiled. And every time entropy increases, more opportunities are lost to stop the avalanche of disorder that sometimes seems ready to swallow us all. Left to its own accord, everything—including opportunity—dissipates. Only a purposeful infusion of energy can breathe it back to life.

In some cases, of course, combating entropy is hardly worth the effort. Recent research has revealed, for example, that local attempts to prevent the erosion of beaches are only hastening the effect. It would not be worth the price to try to replace my dying tooth with a living one, it would probably be impossible. Better to root out the root, put on a cap, and call it quits. Some relationships (between countries or people) are simply not worth saving, and it is surely not worth the energy it would take to try to stop my face from wrinkling or my hair from turning gray. Better to put my energies and resources into something more lasting and useful. Besides, some results of entropy form the foundation for new life—like the rotting wood that becomes the groundspring for new trees.

Perhaps the biggest mistake people make in trying to create order out of chaos is to do it in one fell swoop—like Athena bursting out of Zeus's head fully armed and full grown. The truth is that large-scale cosmic order is built on small-scale local order. Stars don't appear out of nowhere, but accumulate by collecting clumps of matter over millions of years. You couldn't compose a poem if you didn't have letters and words. Social or economic order can't

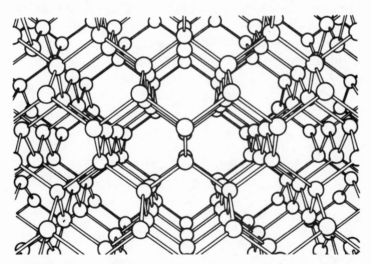

Carbon atoms arranged in a diamond lattice; diamonds are the ultimate in order.

be imposed on chaos, but must be built up by many small simultaneous efforts over long periods of time. A person can evolve from a random collection of atoms in a series of small and often serendipitous steps: atoms to molecules to more complex molecules to simple organisms and so on. But for a person to emerge directly from a random collection of atoms would take more energy than is available in the entire universe.

Many of the most ordered things around us—crystals, flocks of birds, schools of fish, the accumulation of individual cells that make up the human brain—grow upward, so to speak, from the confluence of many local events. Since most complex forms arise without control from above, "anarchy" might be a more appropriate word than "hierarchy" to describe their internal structure—as M.I.T. metallurgist Cyril Stanley Smith points out in his essay *Structural Hierarchy in Science, Art, and History.* He also points out that it "would be good to avoid both terms for they are overloaded with political emotion." Still, it does seem strangely counterintuitive (and rather comforting) to learn that neither large-scale change nor large-scale order requires the intervention of totalitarian-type controllers. Change, like order, tends to move "structurally upward," writes Smith, because parts can adjust faster than wholes. When enough of these local parts transform, they "will find each other and jointly respond by reformation when an appropriate nucleus of a better superstructure appears. This is the way ideas spread and topple governments."

16. SMALL DIFFERENCES

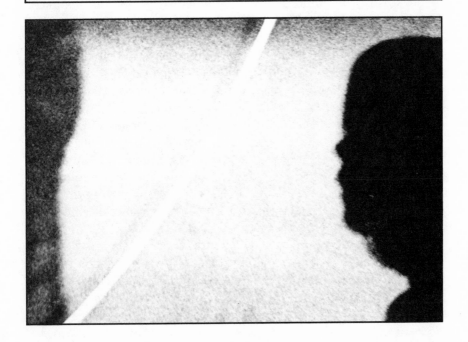

Blobs, spots, specks, smudges, cracks, defects, mistakes, accidents, exceptions and irregularities are the windows to other worlds.
—Artist's statement, by Bob Miller of San Francisco

This sentence floated in front of my mind as I contemplated an article on the latest findings from the sun. Researchers at the University of Arizona, it said, had discovered that the sun was not precisely round, but rather .0017 percent fatter at the equator than at the poles. And yet this almost imperceptible irregularity—

if true—was about to topple the entire Einsteinian foundations of physics, changing not only our understanding of gravity, but also of the structure of space itself. How could it be?

Of course, the road to scientific insight has always been paved with small exceptions and irregularities. The major guideposts and turning points often seem to be insignificant blobs, spots, and specks. The locks to the most impenetrable doors sometimes seem to be sprung by minor, serendipitous discoveries or mistakes. There are times when it all seems like one giant stumble.*

The existence of antimatter was first signaled when physicist Paul Dirac stumbled upon an errant minus sign in an equation. The principle of the electric motor was discovered during a classroom demonstration when high school teacher Hans Christian Oersted happened to notice an unexpected deflection of a current-carrying wire when it moved through a magnetic field. The planet Neptune was found as astronomers tried to account for a small irregularity in the orbit of its neighbor, Uranus. The remnants of the primal Big Bang—radio signals ringing throughout the universe as still "visible" evidence of the first moments of creation—were first picked up as so much "static" on a Bell Labs radio telescope. And so on.

It's amazing how many fundamental theories have hinged on minor anomalies of fact. Einstein's theory of general relativity—now brought into question by a slight squashing of the sun—was proved in the first place by the most menial kinds of evidence: a deflection of starlight passing near the sun by a mere 1.75 seconds of arc (less than a thirtieth of a degree); a small reddish shift in the spectrum of light in a strong gravitational field. How could such meager evidence "prove" something as cosmic as the curvature of space?

One obvious reason that small differences make a big difference in scientific understanding is that small differences can make big differences in nature. The difference of a single electron in an atom's outer shell spells the difference between sodium—one of the most chemically active metals—and neon, a chemically inac-

---

*Arthur Koestler aptly titled his history of scientists' efforts to understand the basic laws of planetary motion *The Sleepwalkers*.

tive gas. A minute change in the wavelength of light turns blue to violet, and allows violet, but not ultraviolet, to pass through glass. If the "strong" force were a little weaker, or the electric force a little stronger, atoms—and therefore matter as we know it—would probably not exist. Indeed, the existence of matter at all is

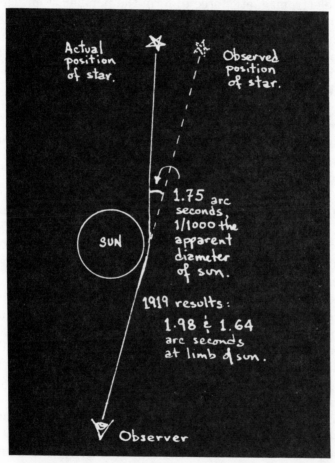

based on what must have been a small imbalance of matter and antimatter sometime near the earliest moments of the universe: particles of matter and antimatter annihilate each other in a burst of energy when they meet; if the universe had been created with equal parts of both, then pairs of particles and antiparticles would have annihilated themselves long ago, leaving us with nothing but radiation.

(Of course, what you mean by "small" depends entirely on how you choose to use the word. You might say that a single electron, for example, is a big difference simply because of the, well, because of the big difference it makes to an atom. And there is some flexibility in just how "strong" the strong force has to be in relation to the electromagnetic force to form the familiar atom. However, I'm using "small difference" to mean what most people consider to be small differences in their everyday lives. A matter of a single electron one way or another is certainly a small difference by that definition. The fact that these small differences often turn out to be big differences is, of course, the point of this chapter.)

Living things are even more profoundly affected by small differences. If the earth had settled into an orbit a little closer to the sun, temperatures would have been so hot that organic molecules could not stay together, but would continually fly apart. If the orbit had been a little farther from the sun, colder temperatures would have frozen opportunities for life into immobility. A tiny alteration in the structure of DNA can mark the difference between brown eyes and blue, between sickness and health, between extinction or survival of a species. One geneticist estimated that the difference between a mild virus and a killer can be as small as three atoms out of more than five million.

These minor differences between major groups carry endless fascination. Annie Dillard points out, for example, that the difference between the lifeblood of plants and people is but one atom: chlorophyll is made up of 136 atoms of hydrogen, carbon, oxygen, and nitrogen arranged in a ring around a single atom of magnesium; hemoglobin (blood) is made up of 136 atoms of hydrogen, carbon, oxygen, and nitrogen arranged in a ring around a single atom of iron. The difference between humans and African great apes—when you consider amino acids, the building blocks of genes—is less than one percent for those genes that have been studied.

That the list of small differences that make big differences is large is hardly surprising. After all, everyone knows how some small irritation can ruin an otherwise pleasant day, just as an irritating habit can turn you away from people you would otherwise like to be around. The small difference between the images that appear on the retinas of your two eyes adds up to impressions of great depth. One percentage point can alter the outcome of a presidential election—which may make a big difference, or none at all. A single shot at Sarajevo changed the history of the world. A small matter like squeezing the toothpaste in the middle of the tube has been known to become a major issue in a shaky marriage.

Yet it's equally obvious that some small differences make much more difference than others. If every small difference threw things askew, then the universe would be hopelessly unstable—and the forms of life on earth would change abruptly every time you turned around. In truth, most small differences don't make much of a difference. Slight changes in temperature, in the shape of your nose, in the way a sentence or a species is constructed, can hardly matter at all. But if the small difference pops up at a crucial point, then it can make all the difference. The difference between 98.6 degrees F. and 106 degrees F. is both small and potentially lethal.

Sometimes the all-important small difference takes the form of what might otherwise seem like unimportant icing on the cake, when in reality it is the essential ingredient that *makes* the cake. The difference between an amateur flutist and Jean-Pierre Rampal might not even be noticeable to the eye. But it would be huge to the ear. The same is true of the last millisecond of time that turns the runner into a champion, the almost imperceptible turn of phrase that gives power to the poem, or the embellishments of style and grace that turn a good dancer into a great one. As philosopher Yehoshua Bar-Hillel said: "The step from not being able to do something at all to being able to do it a little bit is much smaller than the next step—being able to do it well."

Small differences usually make big differences on the strength of their connections. Where one squeezes the toothpaste hardly matters in an otherwise solid relationship, just as a shot in the dark in a poor neighborhood hardly changes the world. A rock doesn't start a landslide unless a whole mountain of rocks is somewhat shaky and ready to fall. These kinds of small differences are triggers. The guns they are connected to must be loaded and cocked for large effects to occur: the combination of genes that sets off a cancerous growth, or the push on the first domino that knocks down the row; the ice crystals that set off a chain of events that culminates in a hurricane, or the small crack beneath the surface of the earth that finally causes Mount St. Helens to blow its stack.

They are what Weisskopf, for one, calls "amplification effects," and
the amplification is particularly powerful when bolstered by such
properties of living systems as reproduction and natural selection.
Shine X rays on salt crystals, he says, and you may change a few
molecules, but you still have a salt crystal. Shine X rays on bacteria and you may alter those very molecules that affect the organism's ability to reproduce. "That is the crux of creation: a small
change in DNA is amplified until it makes a very big difference in a
living being."

(These amplification effects also can help to explain how the
rich get richer and the poor get poorer—almost inexorably. When
you lose your job, you also lose your health insurance, your self-
esteem, and—depending upon the kind of job you had—your sec-
retary, health club membership, Xerox and telephone privileges,
vacations, free travel, and so on. A salary can turn out to be a
relatively small factor in the equation because the effects of having
a job—or not having one—are so vastly amplified. In the same
way, most of the poor in this country are young children. Even
before they are born, they often suffer from insufficient nutrition
and health care; as babies, they lack proper immunizations and
wholesome food; as children, they lack adequate preparation for
school. Since all these effects are amplified at every stage, it is no
wonder that in some areas children are falling like dominoes.)

Unfortunately, it is not always easy to know which things are
connected—and how. That is why earthquake prediction, for example, is so difficult. Seismologists have yet to discover exactly
which of the myriads of strains and stresses present near a fault
will cause two plates of the earth's crust, locked together by friction for years, to lurch apart in an earthquake. So many small
things seem to be important. The same is true of human psychology. No one can predict which combinations of small events will
erupt in a murder or suicide.

Some small differences, however, are connected in such a way
that the results are literally explosive. If you knock down one domino in a row, the chain of dominoes passes along the effect and at
the end of the line you have one knocked-down domino. But say
you have a series of dominoes set up in such a way that the first
domino knocks down two dominoes, and those two knock down
four, and those four knock down eight, and so on. Then you have

what is known as a "nonlinear sequence of events." The effect is not commensurate with the cause—any more than the shot at Sarajevo was commensurate with a global war. What you have, in effect, is an explosion.

My favorite example of how small numbers can add up when they are connected in this way is the familiar story of the fellow who invented chess. The king of the realm, so the story goes, was so pleased with the invention that he offered the inventor any prize he could name. The prize the inventor wanted was a quantity of grain, to be computed as follows: on the first square of the chessboard, the king would put one grain of wheat; on the second square, two grains of wheat; on the third square, four grains; on the fourth square, eight grains; on the fifth square, sixteen grains . . . and so on, doubling the number of grains of wheat for each of the chessboard's sixty-four squares.

How much wheat did the inventor receive? Approximately as much as all the grain that has ever been grown on the North American continent.

Now, doubling something once or twice or even three times doesn't necessarily make a big difference. But if you double anything—no matter how small—enough times (even a relatively small number of times), you always end up with a huge amount. Because doublings are connected like dominoes; each additional doubling is connected to all the doublings before. For this reason, doubling the thickness of a piece of tissue paper just fifty times adds up to enough mileage to reach the moon and back seventeen times. And you can't double any kind of paper more than seven times. If you don't believe it, pick a number (or some paper) and start doubling.

What is even more astounding about the powers of these kinds of "exponential" connections, however, is that people tend not to believe them. They rarely even perceive them.

Several years ago, a Dr. Albert Bartlett described this situation vividly in the *American Journal of Physics*. Bartlett took a population of bacteria doubling inside a Coke bottle once per minute. They started at eleven A.M., and by noon the bottle was full. What time would it be, Bartlett asked, when even the most foresighted of the bacteria realized that they were running out of room? The answer: 11:58 A.M. And even then, the bottle would still

be *three-quarters empty*—so they would have to have been very
farsighted bacteria indeed. At 11:59 A.M., the bottle would still be
half full, or half empty, depending upon your point of view. No
doubt, said Bartlett, the presidents of the bacteria bottle compa-
nies would be running around Bacterialand assuring everyone that
there was no reason to limit the growth rate because, after all,
there was more room still left than had ever been used in the population's entire history. And then—just suppose—they mounted an
extensive effort to explore for new space offshore and, lo and behold, found three new Coke bottles! All the space-starved bacteria
would breathe a long sigh of relief. But how much time would they
have before they were out of room again? Answer: two more minute (two more "doubling times").

Doublings add up fast.

The white line on this plastic is an exponential curve. An exponential curve describes the behavior of many kinds of growth, including population and bank interest.

POPULATION OF THE WORLD

Doublings add up fast.

Bartlett wrote that people simply didn't believe him when he pointed out that at current growth rates we would run out of oil and coal just as the bacteria were running out of bottle room. These people knowingly argued that there was more oil left sitting under the sea and under America's national forests than had ever been burned on the planet. But all that means is one "doubling time"—which at a 7 percent growth rate is just ten years. Even if they found three times that much (three new Coke bottles), it would only give us twenty additional years. The only way to stop the impending energy crisis (or population crisis or whatever) is to drastically slow the growth rate. But a mere 7 percent seems so small! No wonder people are surprised to find their incomes effectively cut in half every ten years when the inflation rate is only 7 percent. At 10 or 15 percent, of course, it doubles much faster. If you take 15 percent of $80,000 five times, you'll see right away why you now need $160,000 to buy the house that $80,000 bought five years ago.

This exponential accumulation of small differences explains everything from nuclear explosions and population explosions to compound interest and avalanches. Yet few people know about it, children aren't taught about it, and even those people who understand how it works have a hard time believing it. It may not even *be* believable in an intuitive sense. It may be simply beyond our powers of perception.

I know of at least two reasons this might be true. One is that often small differences are simply too small to take notice of. Most people don't notice when the department store charge account adds 1.5 percent interest to their balance each month, and it's hard to get excited when you read that the world population is increasing at a rate of 1.8 percent—never mind that this adds up to another billion people in a minuscule period of time. The differences themselves seem small. It's the *connections* between the differences that make them add up—the total pattern—and it's the total pattern that we can't perceive.

An instructive (although somewhat different) example comes from a fable reportedly told by Jorge Luis Borges. It concerns a somewhat mad prince who built columns throughout a vast park. Each one was a beautiful bright red. And each was just like the first. As you passed along, you would see dozens and dozens of them, all the same. And then all of a sudden after a week or so of traveling you would realize that the bright red columns had mysteriously metamorphosed into pure white pillars. How could it be? The answer is that the difference in color between one pillar and the next was too small to discern. But the accumulated difference was enough to change bright red to white. This is similar to British Prime Minister Edmund Burke's observation in the late eighteenth century that "though no man can draw a stroke between the confines of day and night, yet light and darkness are upon the whole tolerably distinguishable."

The second reason we often don't see some of the explosions

that are taking place right in front of our eyes has to do with the
ways our eyes and ears work. Our nervous systems are set up in
such a way that we often perceive small differences when there are
in fact big differences. The sun, my friend the physicist tells me, is
about one hundred thousand times as bright as the moon, and yet
we perceive it as only about twelve times as bright. Hearing works
the same way. It is as if loudness increased like the numbers of
grains on the chessboard, while our ears only registered increases
that corresponded to the number of squares. This kind of "logarithmic" scale is essential to accommodate a broad range of sights
or sounds—just as the logarithmic Richter scale is used to register
a broad range of earthquakes. But it is also innately deceiving. An
earthquake that measures 4 on the Richter scale is ten times more
powerful than one that measures 3. One that measures 5 is only ten
times more powerful than one that measures 4—but one hundred
times more powerful than the 3. Our eyes and ears are Richter
scales, and the sounds and sights around us increase in intensity
like earthquakes. So it would not be surprising if our imaginations
behaved like Richter scales in response to the increasing intensity
of political and social earthquakes of the world too.

I remember the first time I realized this innate inability to perceive large numbers and amounts. I was an editor on the copy desk
of a newspaper, and considered a typo that changed a million into a
billion a relatively minor affair. After all, a million was a lot—
a billion a whole lot more. Like most people, I intuitively felt that a
billion was about as much larger than a million than a million was
larger than a thousand. In each case, the former is a thousand
times larger than the latter, and in each case, all you add is three
little zeros.

But if you start with a millimeter (the thickness of a dime) and
you multiply by a thousand, you have a meter—about a yard. If
you multiply that meter by a thousand, however, you have a kilo-
meter—about two thirds of a mile. The difference between one
millimeter and one thousand millimeters—the first increase—is
999 units. The difference between one thousand and one million—
the second increase—is 999,000 units. And the difference between
a million and a billion—those three little zeros—is *999 million*
units of whatever you happen to be counting. This tends to make
you notice little things like the fact that child nutrition programs

are always funded in millions, while corporate bail-outs and defense boondoggles always add up to billions.

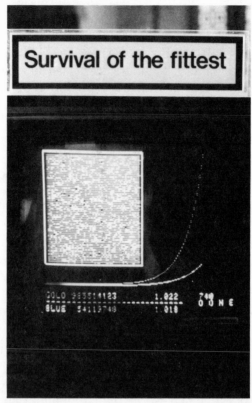

The curve that shoots almost straight upward is growing only four tenths of a percentage point faster than the other curve. This small difference allows "gold" to completely overwhelm "blue."

Two areas where small differences can (and have) added up to particularly dramatic results are evolution, and errors. In evolution, the strength of these connections can turn a small genetic advantage that gives one species an imperceptible edge over another into the difference between survival and extinction. There is an exhibit in The Exploratorium called Survival of the Fittest which is essentially a computer game that allows you to "race" species of colored blips growing at a different rates. If you let one grow at a rate of .18 percent and the other grow at .20 percent (a difference of .02), you can easily see how a favorable trait could wipe out all competition in a geological blink of an eye.

When it comes to errors, we have to remember that everything people believe and perceive is based on long chains of inferences and assumptions.* One weak link in the chain can completely change the conclusions. You can easily misread whole lives and whole histories if you fail to take into account one or two minor—but critical—facts. Just as you can completely misread a sentence if one crucial word is out of place. This is especially true when we speculate about things we know little about—like evolution, or the origins of the universe, or the properties of subatomic particles. "When we speculate about matter under conditions very different from those on Earth," says Weisskopf, "we are bound to make many mistakes. The shortcomings of our understanding are then magnified vastly. We are forced to extrapolate our experience; a small error or misinterpretation can lead us to completely wrong conclusions."

So is the new research on the shape of the sun a crack in the foundation of physics, or isn't it? "Theories like relativity don't depend on one experiment," says Weisskopf. "If you discover one fact that seems out of place, then probably that fact is wrong."

On the other hand, the goal of good science is to know which small, out-of-place bits of fact are critical pieces in the puzzle, which tiny irregularities are harbingers of vast, undiscovered laws. The differences between Einstein's laws and Newton's laws are truly very small—all but imperceptible, for that matter, except at speeds close to the velocity of light. You never notice that time slows down on a trip across the country, or that everyone in the plane gets slightly more massive during the flight. You could never notice the difference between gravity as a Newtonian force and as an Einsteinian curvature of space. Yet the significance of those small differences was enormous. As Richard Feynman points out, Einstein's laws make Newton's laws only a little bit wrong, but "philosophically, we are completely wrong," with Newton's laws. "This is very peculiar thing about the philosophy, or the

*See "Seeing Things."

ideas, behind the laws. Even a very small effect sometimes requires profound changes in our ideas."

In the end, it was the small differences that Darwin noticed in
species of turtles and birds and iguanas that led him down the path
toward evolutionary theory in the first place: the island-by-island
variations in beak and shell and color. Even today, confidence in
evolution is continually bolstered by such small irregularities as
Stephen Jay Gould's now well-known Panda's Thumb—a thumb
that is really an appendage jury-rigged out of an overgrown wrist-
bone. Nearly perfect design is not good evidence for evolution, he
argues, because it would more likely be the handiwork of a nearly
perfect Creator. Rather, he says, "odd arrangements and funny
solutions are the proof of evolution—paths that a sensible God
would never tread but that a natural process, constrained by history, follows perforce."

So even if it doesn't topple relativity, the occasional odd signal
like the recent soundings from the sun can serve as the chinks in
the armor of our complacency—the glitch in the works that jars us
out of our accustomed ways of looking at things. The trick is to pay
attention to all the small differences—at least for a while. The little, out-of-place signals from a child that show he or she is sick or
distressed; the onset of small, strange behaviors that can mark the
beginnings of a major psychological disturbance; the slight change
in temperature that may or may not be the harbinger of a global
change in environment. Most of the time, small differences really
are small differences. Once they're understood, you can relax
about them.

Big differences, when you think about it, are much more difficult to notice. You don't notice the motion of the earth even though
it is spinning around its center at 1,600 kilometers per second, and
around the sun at 110,000 kilometers per second. You don't notice
the flow of your own blood or the activities of your own cells. Major
social and economic trends often pass us invisibly, because they
build up so slowly. You can't feel the speed of your 747 even when
it's whizzing along at five hundred miles per hour.

Sometimes, you need to hit an air pocket before you know that
you're flying. Perhaps the biggest thing that small differences can
do is open our eyes to the presence of larger, unexpected truths.

INDEX

absolutism, 127–129, 140, 145, 146, 154
abstraction, 168–169, 175, 282
Abzug, Bella, 224
acceleration, 140, 143, 147–151, 202
  gravity and, 147–149, 295
accidents, 316
action and reaction, 86–89, 103
adaptation, evolutionary, 318
Advice to a Young Scientist (Medawar), 20, 63, 206
aesthetic of science, 219–231
air resistance, 46
Alice in Wonderland (Carroll), 291
Ames, Adelbert, 70
amplification effects, 339
anomalies, 329–342
antimatter, 171, 222, 256, 330, 331
Aristarchus, 64, 298
Aristotle, 19, 69, 80, 185, 186–187, 258, 301, 302, 303, 318
Armstrong, Neil, 32
art:
  cultural context of, 224
  science and, 220–222, 224
Ascent of Man, The (Bronowski), 20, 29
Asimov, Isaac, 20, 33, 159–160, 199, 308
atomic nucleus, 43–44, 184
atoms, 109, 110–113, 176, 331, 332
  as concept through history, 160

  harmonics of, 271
  nucleus of, 43–44, 189
  size and shape of, 295

Bacon, Francis, 157
Baker, Adolph, 20, 242–243, 250
Bar-Hillel, Yehoshua, 333
Barnett, Lincoln, 15, 20, 23, 57, 58, 74, 82, 131, 136
Bartlett, Albert, 335–337
beauty:
  as order, 319
  in science, 225, 228–230
bees, 249
behavior patterns, 282–283
Berkeley, George, 23
Berman, Zeke, 50
Big Bang Theory, 17, 52, 135, 164, 169, 173, 325, 330
Binet, Alfred, 204
Biography of Physics (Gamow), 18, 55, 203
Biography of the Earth (Gamow), 18
biological clocks, 134
biology, 146, 168
black holes, 27, 28, 91, 101, 129, 140, 151–152, 153, 181, 285
Bloch, Felix, 209–210
Bohm, David, 52–53, 75, 182
Bohr, Niels, 18, 76, 112, 155, 175, 203, 204, 210–211, 213–214, 222, 225
Bondi, Hermann, 195

ABOUT THE AUTHOR

K. C. Cole received her bachelor's degree from Barnard College and has had articles published in *The New York Times*, *Newsday*, *The Washington Post*, and other newspapers and magazines. She has written three science books for The Exploratorium in San Francisco, entitled *Vision: In the Eye of the Beholder*, *Facets of Light*, and *Order in the Universe*. She currently writes a highly acclaimed monthly column for *Discover* magazine and lives in New York with her husband, Peter Janssen, and two children.